"As environments are reverse engineered to match the spreadsheets and management platforms in which they are tallied, the environmental politics of data control, organization, and proliferation will hugely influence ecologies and politics going forward. By putting that insight front and center, Goldstein and Nost assemble a sweeping set of essays that gaze into the sometimes-disturbing future of the planet."
—Paul Robbins, author of *Political Ecology: A Critical Introduction*

"This volume contributes to the growing discourses around political ecological work on data and the infrastructures that sustain, produce, and exchange them. The volume is startling in both its depth and breadth of engagement with timely and important topics; it marks a significant contribution to a growing field."
—Jim Thatcher, author of *Thinking Big Data in Geography:
New Regimes, New Research*

"Throughout, the reader is plunged into the complexities of digital systems, the environments they monitor and conserve, and the limits to their governance and oversight across a variety of places and scales and sovereignties. What emerges is resolutely *not* an endorsement of further digitalization of nature but a recognition that digitalization is perhaps yet another set of processes in which nature is actively produced."
—Matthew W. Wilson, author of *New Lines:
Critical GIS and the Trouble of the Map*

"In accelerating ways, environmental politics are data politics. This powerful book shows what this looks like in different settings and at different scales, persuasively calling for a new subfield focused on the political ecology of data. Extending from prior work on the delimitations and politics of environmental science, the collection draws out what environmental data can help us see, what it cuts out, and how environmental data production itself is both polluting and weighted by commercial interests."
—Kim Fortun, author of *Advocacy after Bhopal:
Environmentalism, Disaster, New Global Orders*

The Nature of Data

The Nature of Data

Infrastructures, Environments, Politics

Edited by JENNY GOLDSTEIN and ERIC NOST

University of Nebraska Press

LINCOLN

CONTENTS

ILLUSTRATIONS

TABLES

The Nature of Data

Introduction

INFRASTRUCTURING ENVIRONMENTAL DATA

Jenny Goldstein and Eric Nost

Our relationships with nature are increasingly digitized. When we look at some of the most pressing issues in environmental politics today, such as equitable adaptation to climate change, control over access to natural resources, and biodiversity loss, it is hard to avoid data technologies. Big data, artificial intelligence, and data dashboards all promise revolutionary advances in the speed and scale at which governments, corporations, conservationists, and even individuals can respond to such challenges.

For example, sensors embedded across Louisiana's sinking coast are providing fine-grain inputs for that U.S. state's climate change adaptation plan, ultimately informing decisions about which communities to invest in and which to abandon.[1] The Nature Conservancy has integrated groundwater data with citizen science observations in California's Central Valley to more precisely select the farmers it will pay for supplying "pop-up" habitat for migratory birds.[2] Meanwhile, the San Francisco–based start-up company Planet has launched hundreds of small satellites to collect high-resolution imagery of human changes to the environment such as oil pads, clear-cut forests, and mining activity across the globe.[3] The firm's goal is to index Earth's surface much in the way Google made it possible to query the internet, making economic activity and environmental impact searchable and findable for investors, researchers, and activists alike. Data technologies such as those Planet is building may quicken and intensify the transformation of nature into capital and reinforce our knowing it as a discrete thing rather than as a web of relationships. Yet as a platform that also purports to serve the aims of conservation watchdog groups, Planet forces us to consider how data may contribute to resource and environmental justice struggles. Who actually stands to benefit from data's use?

The terrain of these digital environmental politics is not limited to the Global North.[4] Throughout the developing world, international aid organizations, state development programs, and transnational nongovernmental organizations (NGOs) are employing smartphone apps, algorithms, and cryptocurrency to intervene in long-standing environmental issues ranging from ecosystem degradation to community land conflicts. The UN World Food Programme is deploying blockchain to give refugees in the Middle East access to food entitlements.[5] In Vietnam, USAID has rolled out financial e-transfer smartphone apps that replace cash transfers to households enrolled in a program that provides payments for environmental services.[6] Donors in the United States check whether individual trees planted on their behalf in Madagascar are still alive using QR codes, geolocation tags, and drone-based verification photos.[7] In the absence of agricultural extension workers, farmers in Malawi use video consoles and WhatsApp to share information on managing invasive pest outbreaks.[8]

In a shift away from long-standing state control over natural resource data, forest communities, companies, and NGOs now track deforestation activity across the tropics through remote sensing–based platforms that are publicly accessible and rely in part on open-source hardware and data layers.[9] Humanitarian OpenStreetMap, a volunteer organization, gathers similar sorts of satellite imagery to map areas, such as rural Peru, that are "completely unmapped" and vulnerable to earthquakes, floods, and even the COVID-19 pandemic.[10] Part and parcel of an emphasis on transparency and anticorruption in international aid and investment, these data platforms are being created by development agencies and organizations promoting information disclosure, good governance, inclusive participation, and risk reduction, with clear extensions into the realm of environmental politics.[11]

The digital realm has become a significant site for environmental politics, be it debates about allocating water to birds, plants, or people in California or holding the world's biggest corporations accountable for their contributions to climate change. Whether we look to the Global North or the Global South, to data-sharing platforms or server "farms," environmental knowledge, control, and conflict are being routed through data technologies, and data technologies through nature. "Software is eating the world," as web innovator and venture capitalist Marc Andreesen has said, be it the sectors he was interested in, such as healthcare and education, or the realm of something like habitat conservation.[12] At the heart of USAID's smartphone apps for environmental service payments or Planet's indexing of Earth are data *infrastructures*—hardware devices such as servers and satellite sensors, software such as algorithms and models, and platforms and institutions—which enable people to generate and analyze environmental data.[13] Data infrastructures inform environmental management at local, regional, and global scales, but they themselves are managed, contributions to them are incentivized, and access to them is regulated.

In this volume, contributors explore the making, maintaining, and effects of the data infrastructures through which conservationists, communities, and corporations—as well as development practitioners, scientists, and state planners—generate actionable, yet uneven, knowledge about the environment. We bring together scholars from disciplines including geography, anthropology, ecology, information science, and science and technology studies (STS) to advance a *political ecology of data*. A political ecology of data begins from the premise that we cannot fully understand the current conjuncture in critical environmental politics without understanding the role of data platforms, devices, standards, and institutions. In collectively crafting a political ecology of data, this volume furthers nature-society scholars' ability to understand how new data technologies shape access to and control of the environment, land, and resources. Political ecologists have shown how political economic structures and institutional cultures inform who creates environmental knowledge, whose perspectives are deemed legitimate, and the values and interests characterizing this knowledge.[14] But as STS scholars have argued, knowledge production also relies on sociotechnical infrastructures: proprietary and open-source devices, embodied practices, and social institutions. They are ubiquitous but often invisible, things that we take for granted but are deliberately built and maintained.[15] Starting from research that observes how environmental science generated with tools like geographic information systems (GIS) and remote sensing drives land management policies, we draw on scholars of the digital to further characterize the infrastructures supporting such tools and show how infrastructures themselves are sources of contestation.[16]

We offer two overarching arguments about how data infrastructures mediate socioenvironmental relationships. First, data infrastructures are unevenly produced and consolidated. Corporate and government decision-makers' investments in data sources and technologies often deepen asymmetries in whose expertise is valued by rerouting knowledge production and application through computer models and other software. At the same time, these data infrastructures are in constant need of both ideological justification and technical maintenance, and they are open to retooling by critical alternatives. Second, data infrastructures have become a key dimension of socioecological flows. Data infrastructures now saturate urban and rural places by mediating flows of cars and carbon, water and wildlife and thus have the potential to remake material environments.

But why infrastructure, and what do we mean by it? Infrastructure is useful as both a topical focus and a conceptual lens. First, we highlight material data infrastructures—the hardware, software, and institutions we mentioned—and their governance. Our relationships with nature may be increasingly digital, but that does not mean they are immaterial. Like other political ecologists who have studied infrastructure, whether dams, canals, energy grids, pipes, or wires, we believe that turning our attention to the nuts and bolts of the digital realm helps shed light on environmental politics.[17] For instance, researchers have noted that the data centers where video streaming and even environmental conservation data processing now take place, in what is known as the cloud, are powered and chilled twenty-four seven by fossil fuel energy, giving them a noticeable footprint in Earth's carbon budget. The companies that run these centers, including Microsoft and Alphabet (Google), have pledged to offset their emissions, adding a new twist to political ecologists' concerns about how carbon governance manifests in specific places across the world.[18] Geographers and others have also assessed how these facilities and other components of digital infrastructure are embedded in land and water, showing that

Goldstein and Nost

they both rely on and reconfigure place-specific socioenvironmental relations.[19] Networked data technologies are directly imbricated in material—that is, political, economic, and environmental—contexts, be they flows of finance and circuits of urban de- or reindustrialization or flows of water.

Second, we see data as infrastructure for decision-making. For us, infrastructure is a concept that has value beyond helping identify the material components of everyday digital life. We are concerned not just with whether this data portal, that server farm, or the newest platform counts as infrastructure. Instead, seeing infrastructure as a practice, or what Susan Leigh Star and others have called "infrastructuring," gives us a way to witness the politics that make data available for use.[20] Seeing data as something to be infrastructured can tell us about the kinds of demands, redistributions, and visibilities—that is, the politics—that arise in and through knowledge production. Whom and what is environmental data for?

This is a view of data in relational terms: data about the environment is not so much collected but made, and it is made meaningful in relation to other data and to specific uses (in Gabrys, Pritchard, and Barratt's terms, as "just good enough," or as Kim Fortun discusses, as "appropriate").[21] This allows us to push back on seductive ideas that are currently emerging: that data is "raw" and is out waiting in the world to be collected, that it can drive decision-making by itself, and that it is inherently useful.[22] With growing interest in "data-driven" environmental governance, data is increasingly understood as a direct input for various kinds of decisions, potentially short-circuiting previous modes of translating science into policy. Our goal, beyond questioning the epistemological assumptions behind the data, is to unravel the sociotechnical *and* political-economic arrangements that enable data to drive governance. Our political ecology of data intends to draw attention to the ways digital technologies inform environmental politics and governance beyond the creation and application of formal scientific expertise.

Organization of the Volume

Data is seen as revolutionary: we are supposed to be living in a "fourth industrial revolution" in which big data and machine learning will drive economic growth and environmental sustainability.[23] To tech innovators, data and its attendant economies are "disruptive," able to reshape supposedly inefficient sectors, including environmental governance.[24] Questions like "Can cryptocurrencies save Indonesia's forest carbon?" or "Can big data stop climate change?" circulate widely. Contributors to this volume explore conservationists', governments', and businesses' promise that disruptive data infrastructures are *the* solutions to governance problems. We contend that revolutionary discourses—applied to environment, data, or both in tandem—should be approached critically.[25] What the rhetoric of revolution, salvation, and disruption fails to get across is that social, political, and economic structures remain durable and influential; that data infrastructures are not made by fiat but are built and practiced and should be designed with social change in mind; and that data infrastructures often reproduce existing extractive modes of relating to land and resources. These are the three points by which we have organized the contributions to this volume.

Sensors, Servers, and Structures

Blaikie and Brookfield famously characterized the then-emerging field of political ecology as "a broadly defined political economy" that provides chains of explanation for environmental change.[26] Scale, political agency and economic structure, and the materiality of socioenvironmental systems remain guiding concerns for the field. Chapters in part 1 of this book make clear that a political ecological approach can be applied to digitally mediated environmental change by following where data flows and considering its material (not just epistemological) relationships. In chapter 1 Graham Pickren looks at a building in Chicago that once served as an industrial bakery and is now, following deindustrialization, being repurposed as a data center. This chapter extends urban political ecological work showing infrastructure as a key site of socionatural metabolism to include digital systems, illustrating that the "revolutionary" data infrastructures we live with today have an emplaced history.

While social media platforms and their attendant infrastructures may represent a new object for political ecology, in chapter 2 Luis Alvarez León looks to a technology long explored by political ecologists—

remote sensing—and reveals three things about the next generation of Earth imagery data infrastructures. First, Global South countries are launching their own satellites and demanding more control over the data these satellites produce, and second, private companies like Planet are stepping into the imagery data industry, promising to create an indexable database of Earth's surface.[27] Third, the implications of this institutional rearrangement between states and private companies gives rise to what Alvarez León calls an emergent satellite ecosystem, characterized by economic neoliberalization.

In chapter 3 Karen Bakker and Max Ritts provide a deep scan of what they call Smart Earth technologies: landscape-embedded sensors and related devices that constitute the data revolution as applied to environmental management. The authors soberly assess the transformational ends posited by these tools, including whether they might enable a real-time and more flexible form of governance.[28] Finally, Anthony Levenda and Zbigniew Grabowski return to the historical and material trajectories of data infrastructures in chapter 4 by tracing how knowledge about nature in the U.S. Pacific Northwest has been (dis)embedded in that region differently since time immemorial. Pointing to Indigenous peoples' demands for data sovereignty, Levenda and Grabowski discuss the nonhuman world itself as a kind of infrastructure for Indigenous knowledge systems, then note how settler-industrial knowledge regimes demand a different, more instrumentalist and extractive infrastructure, one that has culminated in the localization of server centers in the region for global social media platforms like Facebook.[29]

Civic Science and Community-Driven Data

Chapters in part 2 look at case studies of civic science projects that build data collection, storage, and analysis tools to advance socioenvironmental justice. There is a long history of work in political ecology that is concerned with the politics of expertise, but here we place special emphasis on how such politics form around data infrastructures such as platforms for sharing Arctic data (chapter 8) or government websites that make accessible or restrict scientific information about climate change (chapter 10).[30] Authors challenge the notions that more and better data can "save" us and necessarily will lead to more just and equitable socioenvironmental outcomes. Building from political ecological work that highlights knowledge subverting or countering state and corporate goals, and from STS research related to "making and doing," many of these chapters ask scholars and data practitioners to think reflexively about their own contributions to and uses of data infrastructure and what implications their work has for those outside of the academy.[31] They suggest that data technologies can be coproduced with social justice goals.

An important premise here is that it is challenging to think about social and environmental change in the world apart from the tools and objects that increasingly structure our everyday lives. Alternatives designed with justice in mind are not perfect but demand constant reflection on the contested concepts they mobilize, like community and scale (chapter 10). Contributors take hatchets to the extractive, reductionist, ineffective, nonresponsive, and disembodied forms of producing environmental data for knowledge regimes but also attempt to envision data infrastructures that resist these purposes, whether through practices that are "just good enough" for antifracking efforts (chapter 5) or that inform the design of platforms that focus on user perspectives (chapters 6 and 8).

In chapter 5 Jennifer Gabrys and Helen Pritchard describe citizen science efforts to document the impacts of fracking. Their emphasis is on people *as sensors*, and they show us what kinds of data the state, in this case the U.S. Environmental Protection Agency, is willing to accept from its subjects. Advancing the concept of "data citizenships," the authors argue that citizen science should not necessarily aim to replace the state's own data programs. Instead, citizen sensing can lead to new meanings of citizenship and environmental engagement. In chapter 6 M. V. Eitzel and coauthors detail the sociotechnical infrastructures that facilitated collaborative modeling of land use in Zimbabwe among ecologists, computer scientists, residents, and a local nonprofit organization. While interest in these sorts of collaborative knowledge-building projects is growing, it remains unclear how to make them effective and justice-oriented in practice. The authors reflect on the opportunities and challenges in the technical choices

Goldstein and Nost

they made, like the decision to use agent-based modeling because its fine-grain representation of land use was more intuitive for residents.

Irus Braverman expands on citizen science approaches to data collection in chapter 7 in the context of coral reef restoration, exploring tensions between Smart Earth sensor approaches to knowledge development and the kinds of human-collected data necessary for tricky reef restoration projects. In doing so, she charts some of the debates between lay coral reef restorationists and those within academic science. Noor Johnson, Colleen Strawhacker, and Peter Pulsifer show us in chapter 8 how synoptic, large-scale, technocratic Earth systems monitoring projects inform policy, arguing instead in favor of a diversity of community-based data infrastructures. The authors developed online platforms helping different ongoing Indigenous science projects connect to one another, designing them in such a way that the data cannot easily become inputs for someone else's research, as is the case with many large-scale infrastructural efforts. Instead, the platforms purposely create "data friction" in brokering connections between researchers and empowering Indigenous knowledge regimes.[32]

In chapter 9 Patrick Gallagher describes an effort to make land cover data meaningful and actionable for conservationists, as in chapter 6, but with a focus on the affective dimensions of data production. Conservationists in Belize are attempting to implement plans based on ecosystem services modeling and remote sensing. Gallagher is interested in uncovering the diverse politics of conservation embedded in data infrastructures by emphasizing how conservationists bring embodied experiences to bear on their work with abstract data and models, rather than assuming the conservationists align with hegemonic discourses of objectivity, neutrality, and instrumentalism in their use of data technologies.

Finally, in chapter 10 Dawn Walker and coauthors report on work by the Environmental Data and Governance Initiative (EDGI) to "rescue" data that was seen to be at risk of disappearing as the explicitly antiscience Trump administration came to power in the United States. The authors reflect on the fact that the design of EDGI's rescue efforts did not really instantiate the principles at the core of the organization because the whole notion of a rescue divorced this government environmental data from its often unjust production and use. They then envision what alternative data infrastructures might look like, not shying away from the technical affordances that their social goals would require.

Governing Data, Infrastructuring Land and Resources

Data plays an increasingly contradictory role in contemporary resource governance. As authoritarian and antiscience movements take root around the world, elites justify their actions less in claims to objectivity and more in appeals to racism, nationalism, and crude economic determinism. Yet for many conservationists, the response has been to double down on claims to the objectivity and neutrality of data. For others, the current moment demands critique of antiscientism and its rejection of rigor and also presents an opportunity to posit feminist data regimes that better reflect marginalized standpoints.[33] Chapters in part 3 focus on the ways in which states—in tension and in conjunction with civil society, nongovernmental organizations, and the private sector—have designed, funded, and otherwise enabled different kinds of data infrastructures to support environmental management, land control, and ecological research.

In chapter 11 Madeleine Fairbairn and Zenia Kish analyze new "data brokers" that seek to provide digital extension services to farmers in the Global South, looking especially at how these arrangements may intensify the Green Revolution's emphasis on yield over sustainability and livelihoods. The authors are also concerned about data privacy and consent issues, as well as the potential for new digital tools to shift who profits from data and who bears the risk of agriculture. In chapter 12 Hilary Faxon and Jenny Goldstein investigate the rapid introduction of data infrastructure to Myanmar and, more generally, its use in surveilling illicit environmental activities such as logging and wildlife poaching in Southeast Asia. They ask what tensions arise when data that is intended to make knowledge more accessible and transparent to certain publics encounters environments and actors that remain entrenched in governing norms established by authoritarian states. Though it is perhaps too soon to tell what empirical effect the recent proliferation of new data infrastructure has on land relations and land use change, the authors argue that such infrastructure is nevertheless producing new forms of environmental knowledge, which does not necessarily adhere to the tenets of open access and transparency.

Chapter 13, by James Blair, is a case of data infrastructure technopolitics: the use of technology to achieve political goals in the context of resource governance. His contribution discusses a penguin-tracking project launched by the British Falkland Islands Group as part of new offshore drilling efforts. He argues that the state, with help from scientists and in concert with oil firms, uses data infrastructure projects to advance territorial claims. Blair shows how data infrastructure becomes wrapped up in broader political economic struggles that are in turn shaped by colonial legacies.

In chapter 14 Corrine Armistead takes us through the technical details of database structures to demonstrate how they mediate extractive resource governance regimes for hydropower in the U.S. Pacific Northwest. She shows how relational databases may be able to link land, water, and demographic data based on instrumental commonalities among them but cannot represent noninstrumental or Indigenous relationships with land. Echoing contributors in part 2 of the volume focused on data ethics, justice, and alternatives, Armistead goes on to demonstrate how graph databases might better approximate relational understandings of nature-society, without dismissing the idea that data technologies are useful means for us to govern our relationship with the world. Finally, in chapter 15 Cindy Lin offers an account of how data technicians in Indonesia have attempted to automate their own forest mapmaking labor to meet the demands of state bureaucracy. Through ethnographic investigation of the country's One Map and One Data policies, she shows that seeing and knowing nature through remotely accessed data is segregated along racial and class-based lines, even as such mediated access to Indonesia's forest is promoted by the state as a transparent and uncorrupt proxy for ground-truth.

Given the proliferation of data use among various environmental governance actors, there is a politics to why and how software interfaces with our material environment. Software is not just "eating the world," with apolitical implications of decline, material passivity, and technological agency that political ecologists would approach dubiously. In fact, the environmental imbrications of new and expanded data infrastructures dredge up old debates in political ecology over resource access and ecological degradation. While many conservationists, states, and corporations claim that digital tools offer them better insights into nature, it is increasingly clear that data also does other things. Data and its infrastructures are not so much clear, objective windows as they are tinted lenses, where everything about them—from the tint to the frames—shape what and how we see, as well as what gets contested.

Notes

1. Snider, "How Can We Reduce Losses?"
2. Jayachandran, "Using the Airbnb Model."
3. See the Planet website, https://www.planet.com/; see also chapter 2 of this volume.
4. Machen and Nost, "Thinking Algorithmically."
5. WFP, "Blockchain against Hunger."
6. USAID, "USAID Promotes Financial Inclusion."
7. Vyawahare, "Tree-Planting Programs."
8. Parker, "Fighting Africa's Fall Armyworm Invasion."
9. See the Global Forest Watch website, https://www.globalforestwatch.org/; see also Gaworecki, "Shah Selbe."
10. Scoles, "Satellite Data."
11. Gupta, Boas, and Oosterveer, "Transparency."
12. Andreessen, "Why Software Is Eating the World."
13. Bowker, "Biodiversity Datadiversity"; Edwards, *Vast Machine*; Amoore, "Cloud Geographies"; Eghbal, *Roads and Bridges*; Pickren, "'Global Assemblage'"; Srnicek, *Platform Capitalism*; Alvarez León, "Blueprint for Market Construction?"; Mattern, "Maintenance and Care."
14. Fairhead and Leach, *Science, Society and Power*; Goldman, Nadasdy, and Turner, *Knowing Nature*; Lave, "Bridging Political Ecology and STS."
15. Star and Ruhleder, "Steps towards an Ecology of Infrastructure"; Star, "Ethnography of Infrastructure"; Star and Bowker, "How to Infrastructure"; Edwards, *Vast Machine*.
16. For land management politics, see Robbins, "Fixed Categories"; Turner, "Methodological Reflections." For digital geographers, see Dodge and Kitchin, "Flying through Code/Space"; Wilson, "Data Matter(s)"; Leszczynski, "Situating the Geoweb"; Elwood and Leszczynski, "New Spatial Media"; Kinsley, "Matter of 'Virtual' Geographies"; Kitchin and Lauriault, "Towards Critical Data Studies"; Alvarez León, "Property Regimes"; Ash, Kitchin, and Leszczynski, *Digital Geographies*; Thatcher, Eckert, and Shears, *Thinking Big Data*; McLean, *Changing Digital Geographies*. For infrastructure and contestation, see Parks, "Digging into Google Earth"; Gillespie, "Politics of 'Platforms'"; boyd and Craw-

Goldstein and Nost

ford, "Critical Questions"; Starosielski, *Undersea Network*; Srnicek, *Platform Capitalism*; Anand, Gupta, and Appel, *Promise of Infrastructure*.

17. Kaika and Swyngedouw, "Fetishizing the Modern City"; Anand, "Pressure"; Furlong, "Small Technologies, Big Change"; Akhter, "Infrastructure Nation"; Cousins and Newell, "Political–Industrial Ecology"; Barnes, "States of Maintenance"; Carse and Lewis, "Toward a Political Ecology"; Lawhon et al., "Thinking."

18. Bumpus and Liverman, "Accumulation by Decarbonization"; Pasek, "Managing Carbon and Data Flows."

19. Hogan, "Data Flows and Water Woes"; Starosielski, *Undersea Network*; Pickren, "'Global Assemblage'"; Lally, Kay, and Thatcher, "Computational Parasites and Hydropower"; Levenda and Mahmoudi, "Silicon Forest and Server Farms"; Pasek, "Managing Carbon and Data Flows"; Furlong, "Geographies of Infrastructure II."

20. Star and Bowker, "How to Infrastructure."

21. Fortun, "Environmental Information Systems"; Gabrys, Pritchard, and Barratt, "Just Good Enough Data."

22. Gitelman, *Raw Data Is an Oxymoron*.

23. World Economic Forum, "Fourth Industrial Revolution."

24. Bakker and Ritts, "Smart Earth."

25. Walker et al., "Practicing Environmental Data Justice."

26. Blaikie and Brookfield, *Land Degradation and Society*, 17.

27. FNIGC, "First Nations Principles of OCAP®."

28. See also Ritts and Bakker, "New Forms."

29. Rodriguez-Lonebear, "Building a Data Revolution."

30. Lave, "Future of Environmental Expertise."

31. Peluso, "Whose Woods Are These?"; Harris and Hazen, "Power of Maps"; Wylie et al., "Institutions for Civic Technoscience."

32. Edwards, *Vast Machine*.

33. Dillon et al., "Situating Data."

Bibliography

Akhter, Majed. "Infrastructure Nation: State Space, Hegemony, and Hydraulic Regionalism in Pakistan." *Antipode* 47, no. 4 (September 2015): 849–70.

Alvarez León, Luis F. "A Blueprint for Market Construction? Spatial Data Infrastructure(s), Interoperability, and the EU Digital Single Market." *Geoforum* 92 (June 2018): 45–57.

——. "Property Regimes and the Commodification of Geographic Information: An Examination of Google Street View." *Big Data & Society* 3, no. 2 (2016): 1–13.

Amoore, Louise. "Cloud Geographies: Computing, Data, Sovereignty." *Progress in Human Geography* (August 2016): 030913251666214.

Anand, Nikhil. "Pressure: The PoliTechnics of Water Supply in Mumbai." *Cultural Anthropology* 26, no. 4 (November 2011): 542–64.

Anand, Nikhil, Akhil Gupta, and Hannah Appel, eds. *The Promise of Infrastructure*. Durham NC: Duke University Press, 2018.

Andreessen, Marc. "Why Software Is Eating the World." *Wall Street Journal*, August 20, 2011. https://a16z.com/2011/08/20/why-software-is-eating-the-world/.

Ash, James, Rob Kitchin, and Agnieszka Leszczynski. *Digital Geographies*. Thousand Oaks CA: SAGE, 2018.

Bakker, Karen, and Max Ritts. "Smart Earth: A Meta-Review and Implications for Environmental Governance." *Global Environmental Change* 52 (September 2018): 201–11.

Barnes, Jessica. "States of Maintenance: Power, Politics, and Egypt's Irrigation Infrastructure." *Environment and Planning D: Society and Space* 35, no. 1 (February 2017): 146–64.

Blaikie, Piers M., and H. C. Brookfield. *Land Degradation and Society*. London: Routledge, 1991.

Bowker, Geoffrey C. "Biodiversity Datadiversity." *Social Studies of Science* 30, no. 5 (October 2000): 643–83.

boyd, danah, and Kate Crawford. "Critical Questions for Big Data: Provocations for a Cultural, Technological, and Scholarly Phenomenon." *Information, Communication & Society* 15, no. 5 (2012): 662–79.

Bumpus, Adam G., and Diana M. Liverman. "Accumulation by Decarbonization and the Governance of Carbon Offsets." *Economic Geography* 84, no. 2 (April 2008): 127–55.

Carse, Ashley, and Joshua A Lewis. "Toward a Political Ecology of Infrastructure Standards: Or, How to Think about Ships, Waterways, Sediment, and Communities Together." *Environment and Planning A: Economy and Space* 49, no. 1 (January 2017): 9–28.

Cousins, Joshua J., and Joshua P. Newell. "A Political–Industrial Ecology of Water Supply Infrastructure for Los Angeles." *Geoforum* 58 (January 2015): 38–50.

Dillon, Lindsey, Rebecca Lave, Becky Mansfield, Sara Wylie, Nicholas Shapiro, Anita Say Chan, and Michelle Murphy. "Situating Data in a Trumpian Era: The Environmental Data and Governance Initiative." *Annals of the American Association of Geographers* 109, no. 1 (January 2019): 1–11.

Dodge, Martin, and Rob Kitchin. "Flying through Code/Space: The Real Virtuality of Air Travel." *Environment and Planning A: Economy and Space* 36, no. 2 (February 2004): 195–211.

Edwards, Paul N. *A Vast Machine: Computer Models, Climate Data, and the Politics of Global Warming*. Cambridge MA: MIT Press, 2010.

Eghbal, Nadia. *Roads and Bridges: The Unseen Labor behind Our Digital Infrastructure*. New York: Ford Foundation, 2016. https://www.fordfoundation.org/media/2976/roads-and-bridges-the-unseen-labor-behind-our-digital-infrastructure.pdf.

Elwood, Sarah, and Agnieszka Leszczynski. "New Spatial Media, New Knowledge Politics." *Transactions of the Institute of British Geographers* 38, no. 4 (October 2013): 544–59.

Fairhead, James, and Melissa Leach. *Science, Society and Power: Environmental Knowledge and Policy in West Africa and the Caribbean*. Cambridge: Cambridge University Press, 2003.

FNIGC (First Nations Information Governance Centre). "First Nations Principles of OCAP®." 2021. https://fnigc.ca/ocap-training/.

Fortun, Kim. "Environmental Information Systems as Appropriate Technology." *Design Issues* 20, no. 3 (2004): 54–65.

Furlong, Kathryn. "Geographies of Infrastructure II: Concrete, Cloud and Layered (In) Visibilities." *Progress in Human Geography* 45, no. 1 (May 2020): 190–98.

———. "Small Technologies, Big Change: Rethinking Infrastructure through STS and Geography." *Progress in Human Geography* 35, no. 4 (August 2011): 460–82.

Gabrys, Jennifer, Helen Pritchard, and Benjamin Barratt. "Just Good Enough Data: Figuring Data Citizenships through Air Pollution Sensing and Data Stories." *Big Data & Society* 3, no. 2 (December 2016): 205395171667967.

Gaworecki, Mike. "Shah Selbe on How Open Source Technology Is Creating New Opportunities for Conservation." *Mongabay Environmental News*, March 5, 2020. https://news.mongabay.com/2020/03/audio-shahe-selbe-on-how-open-source-technology-is-creating-new-opportunities-for-conservation/.

Gillespie, Tarleton. "The Politics of 'Platforms.'" *New Media & Society* 12, no. 3 (May 2010): 347–64.

Gitelman, Lisa. *Raw Data Is an Oxymoron*. Cambridge MA: MIT Press, 2013.

Goldman, Mara, Paul Nadasdy, and Matthew Turner. *Knowing Nature*. Chicago: University of Chicago Press, 2010.

Gupta, Aarti, Ingrid Boas, and Peter Oosterveer. "Transparency in Global Sustainability Governance: To What Effect?" *Journal of Environmental Policy & Planning* 22, no. 1 (2020): 84–97.

Harris, Leila M., and Helen D. Hazen. "Power of Maps: (Counter) Mapping for Conservation." *ACME: An International Journal for Critical Geographies* (2006): 32.

Hogan, Mél. "Data Flows and Water Woes: The Utah Data Center." *Big Data & Society* 2, no. 2 (December 2015): 205395171559242.

Jayachandran, Seema. "Using the Airbnb Model to Protect the Environment." *New York Times*, December 29, 2017. https://www.nytimes.com/2017/12/29/business/economy/airbnb-protect-environment.html.

Kaika, Maria, and Erik Swyngedouw. "Fetishizing the Modern City: The Phantasmagoria of Urban Technological Networks." *International Journal of Urban and Regional Research* 24, no. 1 (2000): 120–38.

Kinsley, Samuel. "The Matter of 'Virtual' Geographies." *Progress in Human Geography* 38, no. 3 (June 2014): 364–84.

Kitchin, Rob, and Tracey P. Lauriault. "Towards Critical Data Studies: Charting and Unpacking Data Assemblages and Their Work." The Programmable City Working Paper 2, Maynooth University, Maynooth, Ireland, July 2014.

Lally, Nick, Kelly Kay, and Jim Thatcher. "Computational Parasites and Hydropower: A Political Ecology of Bitcoin Mining on the Columbia River." *Environment and Planning E: Nature and Space* (August 2019): 251484861986760.

Lave, Rebecca. "Bridging Political Ecology and STS: A Field Analysis of the Rosgen Wars." *Annals of the Association of American Geographers* 102, no. 2 (March 2012): 366–82.

———. "The Future of Environmental Expertise." *Annals of the Association of American Geographers* 105, no. 2 (March 2015): 244–52.

Lawhon, Mary, David Nilsson, Jonathan Silver, Henrik Ernstson, and Shuaib Lwasa.

"Thinking through Heterogeneous Infrastructure Configurations." *Urban Studies* 55, no. 5 (August 2017): 720–32.

Leszczynski, Agnieszka. "Situating the Geoweb in Political Economy." *Progress in Human Geography* 36, no. 1 (February 2012): 72–89.

Levenda, Anthony M., and Dillon Mahmoudi. "Silicon Forest and Server Farms: The (Urban) Nature of Digital Capitalism in the Pacific Northwest." *Culture Machine* 18 (2019). https://culturemachine.net/vol-18-the-nature-of-data-centers/silicon-forest-and-server-farms/.

Machen, Ruth, and Eric Nost. "Thinking Algorithmically: The Making of Hegemonic Knowledge in Climate Governance." *Transactions of the Institute of British Geographers* 46, no. 1 (March 2021).

Mattern, Shannon. "Maintenance and Care." *Places Journal* (November 2018).

McLean, Jessica. *Changing Digital Geographies: Technologies, Environments and People.* London: Palgrave Macmillan, 2019.

Parker, Stephanie. "Fighting Africa's Fall Armyworm Invasion with Radio Shows and Phone Apps." *Mongabay Environmental News*, October 28, 2019. https://news.mongabay.com/2019/10/fighting-africas-fall-armyworm-invasion-with-radio-shows-and-phone-apps/.

Parks, Lisa. "Digging into Google Earth: An Analysis of 'Crisis in Darfur.'" *Geoforum* 40, no. 4 (July 2009): 535–45.

Pasek, Anne. "Managing Carbon and Data Flows: Fungible Forms of Mediation in the Cloud." *Culture Machine* 18 (2019). https://culturemachine.net/vol-18-the-nature-of-data-centers/managing-carbon/.

Peluso, Nancy Lee. "Whose Woods Are These? Counter-Mapping Forest Territories in Kalimantan, Indonesia." *Antipode* 27, no. 4 (October 1995): 383–406.

Pickren, Graham. "'The Global Assemblage of Digital Flow': Critical Data Studies and the Infrastructures of Computing." *Progress in Human Geography* 42, no. 2 (October 2016): 225–43.

Ritts, Max, and Karen Bakker. "New Forms: Anthropocene Festivals and Experimental Environmental Governance." *Environment and Planning E: Nature and Space* (November 2019).

Robbins, Paul. "Fixed Categories in a Portable Landscape: The Causes and Consequences of Land-Cover Categorization." *Environment and Planning A: Economy and Space* 33, no. 1 (January 2001): 161–79.

Rodriguez-Lonebear, Desi. "Building a Data Revolution in Indian Country." In *Indigenous Data Sovereignty*, vol. 38, edited by Tahu Kukutai and John Taylor, 253–72. Canberra, Australia: ANU Press, 2016.

Scoles, Sarah. "Satellite Data Reveals the Pandemic's Effects from Above." *Wired*, April 9, 2020. https://www.wired.com/story/satellite-data-reveals-the-pandemics-effects-from-above/.

Snider, Natalie Peyronnin. "How Can We Reduce Losses from Coastal Storms? Monitor the Health of Our Coasts." *Growing Returns* (blog), Environmental Defense

Goldstein and Nost

Fund, October 17, 2018. http://blogs.edf.org/growingreturns/2018/10/17/coastal
-storms-climate-change-data-monitoring/.

Srnicek, Nick. *Platform Capitalism*. Cambridge, UK: Polity, 2017.

Star, Susan Leigh. "The Ethnography of Infrastructure." *American Behavioral Scientist* 43,
no. 3 (November 1999): 377–91.

Star, Susan Leigh, and Geoffrey C. Bowker. "How to Infrastructure." In *Handbook of New
Media: Social Shaping and Social Consequences of ICTs*, edited by Leah A. Lievrouw
and Sonia Livingstone, 151–62. London: SAGE, 2002.

Star, Susan Leigh, and Karen Ruhleder. "Steps towards an Ecology of Infrastructure:
Complex Problems in Design and Access for Large-Scale Collaborative Systems."
Proceedings of the 1994 ACM Conference on Computer Supported Cooperative Work
(1994): 253–64.

Starosielski, Nicole. *The Undersea Network*. Durham NC: Duke University Press, 2015.

Thatcher, Jim, Josef Eckert, and Andrew Shears, eds. *Thinking Big Data in Geography: New
Regimes, New Research*. Lincoln: University of Nebraska Press, 2018.

Turner, Matthew D. "Methodological Reflections on the Use of Remote Sensing and
Geographic Information Science in Human Ecological Research." *Human Ecology* 31 (2003): 255–79.

USAID (U.S. Agency for International Development). "USAID Promotes Financial Inclusion While Expanding Vietnam's Forest Protection." February 14, 2019. https://www
.usaid.gov/vietnam/program-updates/feb-2019-usaid-promotes-financial-inclusion
-while-expanding-vietnam-forest-protection.

Vyawahare, Malavika. "Tree-Planting Programs Turn to Tech Solutions to Track Effectiveness." *Mongabay Environmental News*, November 22, 2019. https://news.mongabay
.com/2019/11/tree-planting-programs-turn-to-tech-solutions-to-track-effectiveness/.

Walker, Dawn, Eric Nost, Aaron Lemelin, Rebecca Lave, and Lindsey Dillon. "Practicing Environmental Data Justice: From DataRescue to Data Together." *Geo: Geography and Environment* 5, no. 2 (July 2018): e00061.

WFP (World Food Programme). "Blockchain against Hunger: Harnessing Technology in
Support of Syrian Refugees." May 30, 2017. https://www.wfp.org/news/blockchain
-against-hunger-harnessing-technology-support-syrian-refugees.

Wilson, Matthew W. "Data Matter(s): Legitimacy, Coding, and Qualifications-of-Life."
Environment and Planning D: Society and Space 29, no. 5 (October 2011): 857–72.

World Economic Forum. "Fourth Industrial Revolution for the Earth: Harnessing Artificial
Intelligence for the Earth." January 2018. https://www.pwc.com/gx/en/sustainability
/assets/ai-for-the-earth-jan-2018.pdf.

Wylie, Sara Ann, Kirk Jalbert, Shannon Dosemagen, and Matt Ratto. "Institutions for
Civic Technoscience: How Critical Making Is Transforming Environmental Research."
Information Society 30, no. 2 (March 2014): 116–26.

One

Sensors, Servers, and Structures

ONE

....

Data's Metropolis

THE PHYSICAL FOOTPRINTS OF DATA
CIRCULATION AND MODERN FINANCE

Graham Pickren

We live in a data-driven world. The generation of huge volumes of information about nearly every detail of life by social media applications, "smart" cities, the Internet of Things, and data stored in the cloud, has revolutionized everything from business to social relations to environmental governance. Yet the accumulation and circulation of data are made possible through the built physical infrastructure that underpins it. This chapter targets the networks of data centers, fiber-optic cables, and cell towers that power the transmission of digital data and make the other emergent digital environmental practices discussed in this book possible. Like the factories, railroads, and highways that formed the backbone of the Global North's industrial economy, the infrastructure of computing is now central to twenty-first-century socionatural metabolisms.

As a way to illustrate this convergence, the first section of the chapter takes a broad view of how data in general is produced through infrastructural assemblages of capital, nature, labor, space, and technology. I draw conceptual inspiration from William Cronon's seminal work in *Nature's Metropolis*, which traces how the resources of Chicago's hinterlands (including its prairies, soils, and forests) were transformed into commodities (wheat, timber, meat) tied to urban, regional, and national markets.[1] Technologies such as grain elevators, railroads, and refrigeration enabled these processes of transformation in which the abstraction of tradable units of wheat or beef that the Chicago Board of Trade dealt in became detached from the material experience of crops or animals that producers knew. Likewise, data today is a kind of second nature, an easily fetishized commodity whose relations of production are obscured at the point of their consumption or use.

And as with nineteenth-century commodity markets, unpacking how data is produced reveals processes of uneven geographic development.

In focusing on Chicago's place in the global network of data production and circulation, I provide examples of how the city's manufacturing and industrial building stock has been transformed to serve the needs of the data industry. Buildings where workers once processed checks, baked bread, and printed Sears catalogs now stream Netflix and host servers used for financial trading. These buildings are a kind of witness to how the U.S. economy has changed, but more than that, they are what Shannon Mattern has described as the "bleed points" where the physical and the virtual meet.[2] By exploring these changes in the urban landscape and these bleed points, we not only get a better sense of how data exists in the physical realm but also are better able to conceptualize how data-driven environmental governance drives urban and regional (re)development via the growth of data infrastructure. The circulation of data through the Chicago area is now producing space and binding places together in ways that Cronon's commodities of timber, grain, and meat did in the nineteenth and early twentieth centuries. In thinking about data practices as producing space, we can put digital environmental practices in the historical and geographic context of capitalism's uneven spatial development. The first section of the chapter uses photographs and archival images to help illustrate this context while also bringing the bleed points of the digital age into clearer focus.

I then show how the data centers and fiber-optic cables that have emerged through the adaptive reuse of Chicago's buildings and infrastructure are used for several purposes. Banking and financial services in this city accounted for 10 percent of overall data center demand in 2015.[3] Within this sector, high-frequency trading (HFT) firms have been some of the biggest clients of downtown data center developments.[4] HFT firms use algorithms to execute financial transactions over extremely brief windows of time, creating minuscule profits per trade that add up to significant gains across larger timescales. For these firms, having the fastest algorithm and shortest latency—that is, the fastest connection between trading computers—provides advantages that can be extremely profitable. Therefore, the need for speed has created an "arms race" in the financial world that manifests in the development of specialized data centers, new fiber-optic cable routes, and even microwave towers that can transmit trading information faster than the competition. The arms race to construct HFT networks is related to processes of landscape change that are materially intensive and rapidly evolving but hidden in plain sight. Finally, I provide a brief overview of the role of data centers in uneven geographic development and describe the adaptive reuse of industrial buildings in Chicago for data purposes as an "analog to digital" shift. I conclude by discussing the relationship between data and finance, focusing on the development of specialized financial trading infrastructure through which assets, including those derived from nature, can flow.

The Production of Data and the Production of Space

Data centers have been described as the factories of the twenty-first century.[5] A data center is a facility that contains servers that store and process digital information. When we hear about data being stored in the cloud, that data is really being stored in a data center. But contrary to the ephemeral-sounding term *cloud*, data centers are actually incredibly energy- and capital-intensive infrastructure. Servers use tremendous amounts of electricity, which generates large amounts of heat, which in turn requires extensive investments in cooling systems to keep servers operating. These facilities also need to be connected to fiber-optic cables, which deliver information via beams of light and are the "highway" part of the "information superhighway." In most places fiber-optic cables are buried along the rights-of-way provided by existing road and railroad networks, meaning the pathways of the internet are shaped by previous rounds of development.[6]

What is important to keep in mind here is that an economy based on information, just like one based on manufacturing or agriculture, still requires a built environment through which inputs and outputs circulate. In other words, place still matters in an information economy. For the data industry, taking advantage of the places that have the power capacity, the building stock, the fiber-optic connectivity, and the proximity to both customers and other data centers is often central to its real estate strategy. As this real estate strategy plays out, what is particularly interesting is the way in which infrastructure constructed to meet the needs of a different era is now being repurposed for the data sector. In Chicago's South Loop the former R. R. Donnelley & Sons printing factory, once one of the largest printers in the United States, producing everything from Bibles to Sears catalogs, is now the Lakeside Technology Center, one of the largest data centers in the world and the largest consumer of electricity in the state of Illinois.[7] The eight-story gothic-style building contains vertical shafts formerly used to haul heavy stacks of printed material between floors; these columns are now used to run fiber-optic cabling through the building (which comes in from the railroad spur outside). Heavy floors built to withstand the weight of printing presses are now used to support rack upon rack of server equipment. What was once the pinnacle of the "analog" world of the printed word is now a central node in global financial and telecommunications networks.

Just a few miles south of Lakeside Technology Center is the former home of Schulze Baking Company in the South Side neighborhood of Washington Park. Once famous for its butternut bread, the five-story terra cotta bakery is currently being renovated into the Midway Technology Center, a data center (see figs. 1, 2, and 3). Like the project in the South Loop, the Schulze bakery contains features useful to the data industry. The building also has heavy-load-bearing floors, as well as louvered windows designed to dissipate the heat from bread ovens (or now, servers). And it isn't just the building itself that makes Schulze desirable; it's the neighborhood as a whole. I interviewed a developer working on the Schulze redevelopment project who told me that because the surrounding area had been deindustrialized, and because a large public housing project—the Robert Taylor Homes— had been shuttered in recent decades, the nearby power substations had plenty of idle capacity to meet the data center's needs. For neighborhoods like Washington Park, capital is once again flowing through and sparking surplus value creation, but this swell of investment does not raise as many boats as it used to under previous regimes of accumulation. The neighborhood is prized for its infrastructure but not necessarily for its people. In fact, while the developer of the Midway Technology Center is also from the neighborhood and expressed an interest in having job training for locals at the data center, he told me that the low population density of the neighborhood was an added bonus for building security.

Examples of this adaptive reuse of industrial building stock abound. The former *Chicago Sun-Times* printing facility recently became a 320,000-square-foot data center; a Motorola office building and former television factory in the suburbs was bought by one of the large data center companies; and the once mighty retailer Sears, which has one of the largest commercial real estate portfolios in the country, has created a real estate division tasked with spinning off some of its stores into data center properties.[8] Even the former State Line Generating Plant, a coal-fired power plant just outside Chicago, is now slated to become a data center, as the building's power connections and access to water make it ideal for this purpose.[9]

To be sure, not every data center project involves reusing existing buildings. Many of the large tech companies, like Facebook and Google, focus on building stand-alone, state-of-the-art facilities tailored to their needs. But even in these cases, place still matters. For instance, Facebook, Google, and Microsoft have all built large data centers in the Pacific Northwest in regions with cheap electric power and high-voltage power lines that formerly served the timber and mining industries (see chapter 4). Indeed, it is not just urban areas that are seeing data center capitalization. But the common thread is that across urban adaptive reuse projects and rural developments, there is no blank slate on which the world of data simply emerges. What we see in these anecdotes is the seesaw of capitalist development: decline actually creates conditions ripe for economic growth. As certain industries and regions decline, some of the infrastructure retains its value, thus providing incentives for future investors to seize an opportunity. But more broadly, understanding where the infrastructure of computing is, and why, requires grappling with previous rounds of uneven capitalist development spanning back a century or more to the development of the railroads, the telegraph, and the industrial and political needs of the nineteenth and twentieth centuries. Cycles of boom and bust, tensions between capital's fixity and mobility, and the shifting prominence of a "cognitive–cultural capitalism" vis-à-vis manufactur-

1. The Schulze Baking Company, which operated on Chicago's South Side from 1914 to 2004. The historic building is being turned into a data center. Courtesy of author.

2. Bakers working the conveyor belt at Schulze Baking Company, circa 1920. The new data center will employ significantly fewer workers than the bakery. Fred A. Behmer, for the Jeffrey Manufacturing Company, via Wikimedia Commons.

3. Inside the Schulze Baking Factory in February 2016 in preparation for remodeling into the Midway Technology Center. Courtesy of author.

ing all provide much-needed context in understanding big data and computing today, regardless of how these technologies are applied.[10]

Data centers and fiber-optic cables make up part of the physical infrastructure that powers the storage, processing, and transmission of digital data. This infrastructure is more than just the backbone of the internet—it is also central to the operation of modern finance capital, which increasingly sees the circulation of nature-based assets as central to operations. The following section explores the relationships among data, nature, and finance capital through the lens of computing infrastructure, which serves as the means of financial production for high-frequency trading firms. The twin ascendancy of computing and finance represents a rich site for research. As new enclosures of nature are funneled through data-based financial practices, the infrastructural underpinnings of finance require even more scrutiny.

High-Frequency Trading and the Data Center Arms Race

The processes of creative destruction that have refashioned factories into data centers have likewise generated new relationships between capitalism and nature that are reworking landscapes in profound ways. This "greener" form of capitalism seeks to unleash market forces in order to save, rather than destroy, nature. One prominent thread taken up over the last decade by scholars of neoliberal nature has focused specifically on the financialization of nature, broadly defined as attempts by hedge funds, banks, pension funds, and other financial actors to simplify and repackage the complexities of the natural world into assets that can be traded and profited from. A wetland banking scheme, for example, allows a developer in one location to destroy a wetland in exchange for payments to companies preserving wetlands elsewhere. The financial market is the mechanism by which "conservation" operates.

Digital tools have become central in the formation of these kinds of "nature that capital can see," particularly with regard to how the monetary value of, say, a wetland can be determined.[11] Data centers and related infrastructure are likewise digital tools that are playing a role, albeit indirectly, in facilitating the commodification of nature as the realm of finance turns more toward data-driven practices. In this coming together of data, nature, and finance, place matters in ways that have previously not been recognized. As Bram Büscher and Robert Fletcher note, the development of carbon markets, wetland banking, and other environmental derivatives represent a shift in which the actual material reality of nature is recast as a service that can be traded upon.[12] Nature as a set of services removes place from the ledger, rendering a wetland or other ecosystem an abstraction that can be "protected" by financial actors anywhere around the globe. Those wetland credits can be traded, become collateral, or be packaged into portfolios in various ways, but over time, Büscher and Fletcher argue, the wetland's value becomes increasingly abstracted from the actual use value (or, better yet, the noncapitalist value) of the wetland on the ground.

Missing from this conversation is how financial products move through actual infrastructure in the twenty-first century. Carbon, wetlands, and forests can be converted into abstracted, "virtual" financial credits that have no geography, but the digital ledgers through which these credits circulate *do* have a geography in the form of data centers and fiber-optic cables that serve the financial industry. While Büscher and Fletcher are correct to suggest that capital is entering a new phase in which temporal fixes for capital (i.e., finance) become more significant vis-à-vis spatial fixes (i.e., investments in fixed capital), physical investments are still occurring and reshaping places. Financial centers like Chicago once again serve as instructive examples of how the financialization of nature requires a kind of dialectic, fluctuating between moments of abstraction in which nature is channeled into financial products and moments of fixed capital investment that further enable those financial products to circulate. The development of infrastructure for high-frequency trading is an example of how the moment of financialization generates development in place.

What is the "real," material stuff of finance capital? Approaching this question relationally, Donald MacKenzie has written about the deeply social, embodied practices of open-outcry financial trading done through shouting, hand signals, eye contact, and the physical jockeying for space within stock exchange trading pits.[13] As open-outcry trading has disappeared, electronic forms of trading have taken its place. The physical labor of trading is now the digital labor of algorithms, data centers, and wired and wireless information transmission infrastructure communicating trade positions in increasingly opaque circumstances. But this digital, algorithmic labor is far from immaterial; it enrolls places, people, and resources into action just as the telegraph and open-outcry trading did in previous eras. More than just a technological and social shift, the rise of HFT has also been a geographic one, with an arms race taking place within the data infrastructure world as specialized firms look to set up computing infrastructure in places that offer the absolute shortest route between stock exchanges. Geography is not passive here, but rather is central to the ability of HFT firms to produce surplus value. The following discussion of HFT illustrates the ongoing significance of physical, material relations to the emerging nexus of data, finance, and nature.

Traditional stock markets are concerned with the buying and selling of shares, and the basic innovation of HFT is the devolution of the responsibility for buying and selling to algorithms. While there are several ways that HFT firms make money, one common strategy is for HFT algorithms to exploit the differences between the bid price (what buyers want to pay) and the asking price (what sellers want to be paid) of futures, shares, bonds, currencies, or potentially, carbon permits and wetland credits. For example, "when a share on the Deutsche Bourse [stock exchange] is out of sync with its equivalent futures contract in London or vice versa, HFT computers will simultaneously buy the cheaper one and sell it on the more expensive market."[14] Crucially, because these "spread" differences may exist for minuscule units of time (milliseconds), only algorithms are able to take advantage of price discrepancies, earning a profit of fractions of a penny each time they buy low and sell high. Traditional investors, like pension fund managers, may be looking at price data on a computer screen that is 1.5 seconds (1,500 milliseconds) old, but since algorithms can perform trades in 0.0001 seconds (one-tenth of a millisecond), the human investor doesn't stand a chance.[15] As one critic of HFT put it, "By the time the ordinary investor sees a quote, it's like looking at a star that burned out 50,000 years ago."[16] For this reason, HFT investors often refer to their non-HFT competitors as "whales," "low-hanging fruit," "dumb money," or "dinner."[17]

The essence of HFT is the ability to act on market information faster than one's competitors. HFT firms need three things to function: information about markets (order books from stock exchanges), algorithms (software) to run the trading decisions, and computing infrastructure that can transmit trading information between exchanges (data centers) with the lowest latency, or lag time. Having a faster algorithm (being able to process trades more rapidly than one's competitors) and having the fastest connection between servers across physical space represent the two dimensions of what has been described as an arms race for HFT firms.

In this arms race the initial race was mainly a battle between the algorithms, as firms sought to optimize their computer programs to be faster, smarter, and more efficient than those of their competitors. But in 2010 a firm called Spread Networks announced that it had spent $300 million to lay a new fiber-optic cable between the Lakeside Technology Center in Chicago (the former printing facility discussed earlier) and Nasdaq's servers in Carteret, New Jersey, shaving off three milliseconds from the previous route. Whereas the previous cable meandered because it followed the easiest path along the rights-of-way provided by existing road and rail lines, Spread Networks bought its own rights-of-way, tunneling through Ohio fields and Pennsylvania mountains on a more direct route (and in complete secrecy).[18] By reducing the latency for HFT firms, even by just three milliseconds, Spread Networks could reportedly expect to charge firms $300,000 per month to carry HFT firms' orders through its fiber.[19]

This launched what has been described as the "spread revolution," in which boutique telecommunications infrastructure firms entered into intense competition to reduce latency through the creation of new physical networks between financial centers. The CEO of one network provider stated that "from 2010, the network became more critical than the algo [algorithm]—all of a sudden you had to be in the top five percentile from a speed perspective, no matter how good your algo was. The network trumped the algos."[20] Two other firms—Hibernia Atlantic and Perseus Telecom—sprang into action to lay faster transatlantic cables between New York and London, both at tremendous cost.

Further pushing the envelope are wireless technology companies that use microwave, millimeter wave, and laser, instead of fiber optics, to transmit data. For example, Spread Networks' latency times were quickly beaten by McKay Brothers, who erected a series of twenty microwave towers between data centers in Chicago and New York that allowed HFT firms to make the round-trip in only nine milliseconds (compared with Spread's thirteen milliseconds).[21] Since then towers located on the shortest routes between financial data centers—Chicago to New York, Frankfurt to London—have become so crowded with HFT firms that few to no frequencies are available for use.[22] Transmitting trades via microwave, millimeter wave, or laser is faster than fiber optics, although they carry less data (less bandwidth). However, signals weaken over distance and have to be connected via a whole set of microwave towers that repeat the signal. Thus, for use in transoceanic communication, fiber retains an advantage over wireless technology. Using weather balloons or drones suspended in the air over the oceans to repeat wireless signals likely represents a next phase of the arms race.[23]

This race to be the fastest is pushing revenues down and costs up for HFT firms. As one Deutsche Bank analyst notes, the advanced algorithm technology and ultra-low latency infrastructure being deployed by HFT firms enables trading on price disparities across exchanges at ever faster speeds, meaning the profit from individual trades is getting smaller as the window for arbitrage gets smaller. On top of that, the small niche of companies that assemble the servers and networks, like McKay Brothers, pass on their sunk costs to trading firms and raised their fees significantly between 2010 and 2015.[24] Thus the ability to seize on the tiny geographic and technological differences that is the core of HFT becomes more and more exclusive; true to the "arms race" label, in this game the competitors tend to cancel each other out. In the United States revenues have declined from $7.2 billion in 2009 to $1.3 billion in 2014, and HFT's share of the total trading volume in the United States and Europe appears to have reached its maximum. Between 2005 and 2010 HFT surged from virtually zero to 40 percent of equity trading in Europe and from 20 to 60 percent of trading in the United States. Since the 2008 financial crisis HFT's share of total trading has receded to 35 and 50 percent, respectively.[25]

Conclusions

While HFT appears to have reached its zenith in terms of profitability and market share, computing infrastructure is nonetheless integral to the day-to-day activities of both HFT firms and more traditional financial actors like banks and hedge funds. While it is impossible to discern precisely how, when, and whether nature-based financial assets circulate through these networks because of corporate secrecy, it stands to reason that environmental service trading platforms such as the European Union Emissions Trading Scheme, the London Carbon Exchange, and other market participants rely on these data infrastructures. Furthermore, the production of market information that is fed into the trading algorithms is facilitated by digital tools. In short, in thinking about financialized nature along various points in the commodity network, we see that data infrastructures are crucial to the circulation and production of liquid natural capital.

The incorporation of critical issues like climate change and biodiversity loss into the heart of neoliberal orthodoxy requires that these phenomena be understood as problems of market failure. Within this embrace, only a deepening of market penetration can be put forward to address problems created by markets. There is no alternative. The vast scholarship on neoliberal natures has documented the processes by which nature is abstracted and rendered into a commodity whose value can be recognized by capital. As capitalist surplus value production leans more heavily on financialization, so too do new forms of commodified nature become enrolled in complex financial processes. Both nature and finance, independently and in relation to one another, work as a kind of fix that has become foundational to late neoliberal capitalism. Less has been said, however, about the ways in which the financialization of nature has also required the production of material infrastructures such as data centers, fiber-optic cables, and radio towers through which natural capital circulates. Taken together, the financialization of nature therefore relies on a dialectic between moments of abstraction and moments of fixity, where the production of tradable units of carbon, for example, can animate changes in the physical landscape through the redevelopment of a bread factory into a data center. The financialization of nature flows in and among the spaces of everyday life.

The descriptions here of the intersections between data, nature, and finance get at the novelty of the current moment, but more important, they speak to the longer-term phenomenon of capital's uneven geographic development. Data centers and fiber-optic cables today work to facilitate a *particular* geographic, temporal, and organizational mode of capital accumulation, just as grain elevators, railroads, and trading floors did in Cronon's Chicago. These links to the past serve as pointed reminders that the kinds of material infrastructures being built around data, nature, and finance today will also be devalued and turned over. The data-driven practices described in this volume are indeed new and need to be understood on their own terms, but the path dependencies that made them possible must also be a part of the analysis.

Notes

1. Cronon, *Nature's Metropolis*.
2. Mattern, "Interfacing Urban Intelligence."
3. JLL Research, *Data Center Outlook*, 12.
4. Pletz, "Coresite Buys Downtown Site."
5. Cook, *How Dirty Is Your Data?*
6. Burrington, "How Railroad History Shaped Internet History."
7. Chowdhury, "8 Largest Data Centers."
8. Harley, "QTS Opens Data Center"; see also Pickren, "Factories of the Past"; Sverdlik, "Digital to Convert"; Ryan, "Sears Replaces Retail Stores."
9. Lydersen, "Northwest Indiana Defunct Coal Plant Site."
10. Scott, "Beyond the Creative City."
11. Robertson, "Nature That Capital Can See."
12. Büscher and Fletcher, "Accumulation by Conservation," 14.
13. MacKenzie, "Be Grateful for Drizzle."
14. Onstad, "Lasers, Microwave Deployed."
15. Smith, "Fast Money."
16. Adler, "Raging Bulls."
17. Smith, "Fast Money."
18. Adler, "Raging Bulls."
19. Timms, "Feeding the Need."
20. Timms, "Feeding the Need," 8.
21. Adler, "Raging Bulls."
22. Onstad, "Lasers, Microwave Deployed."
23. Adler, "Raging Bulls."
24. Kaya, "High-Frequency Trading."
25. Kaya, "High-Frequency Trading."

Bibliography

Adler, Jerry. "Raging Bulls: How Wall Street Got Addicted to Light-Speed Trading." *Wired*, August 3, 2012. https://www.wired.com/2012/08/ff_wallstreet_trading/.
Associated Press. "Competing for Data Centers: $1.5 Billion in Tax Breaks, but the Bene-

fits Are Debatable." *Omaha World-Herald*, October 8, 2015. http://www.omaha.com
/money/competing-for-data-centers-billion-in-tax-breaks-but-the/article_8a82805a
-a870-53fe-8910-928aeb4e0961.html.

Burrington, Ingrid. "How Railroad History Shaped Internet History." *Atlantic*, November
24, 2015. https://www.theatlantic.com/technology/archive/2015/11/how-railroad
-history-shaped-internet-history/417414/.

Büscher, Bram, and Robert Fletcher. "Accumulation by Conservation." *New Political
Economy* 20, no. 2 (2015): 273–98.

Chowdhury, Kaushik Roy. "The 8 Largest Data Centers in the World in 2020." Analytics Vidhya, September 2, 2020. https://www.analyticsvidhya.com/blog/2020/09/8
-largest-data-centers-world-2020/.

Cook, Gary. *How Dirty Is Your Data?* Amsterdam: Greenpeace International, 2011. http://
www.greenpeace.org/international/Global/international/publications/climate/2011
/Cool%20IT/dirty-data-report-greenpeace.pdf.

Cronon, William. *Nature's Metropolis: Chicago and the Great West.* New York: W. W.
Norton, 1991.

Harley, Michael. "QTS Opens Data Center at Former Chicago Sun-Times Printing Site."
Data Center Dynamics, July 8, 2016. http://www.datacenterdynamics.com/content
-tracks/design-build/qts-opens-data-center-at-former-chicago-sun-times-printing
-site/96547.fullarticle.

JLL Research. *Data Center Outlook: Strong Demand, Smart Growth.* Chicago: Jones Lang
LaSalle, 2016. https://assets.recenter.tamu.edu/Documents/MktResearch/US-North
-America-Data-Center-Outlook-2016-JLL.pdf.

Kaya, Orçun. "High-Frequency Trading: Reaching the Limits." *Deutsche Bank Research*,
May 24, 2016. https://www.dbresearch.com/PROD/DBR_INTERNET_EN-PROD
/PROD0000000000406105/High-frequency_trading:_Reaching_the_limits.pdf.

Lydersen, Kari. "Northwest Indiana Defunct Coal Plant Site Slated for Massive Data
Center." *Energy News Network*, April 30, 2018. https://energynews.us/2018/04/30
/midwest/northwest-indiana-defunct-coal-plant-site-slated-for-massive-data-center/.

MacKenzie, Donald. "Be Grateful for Drizzle." *London Review of Books* 36, no. 17 (September 2014): 27–30.

Mattern, Shannon. "Interfacing Urban Intelligence." *Places Journal* (April 2014). https://
placesjournal.org/article/interfacing-urban-intelligence/#ref_4.

Onstad, Eric. "Lasers, Microwave Deployed in High-Speed Trading Arms Race." Reu-
ters, May 1, 2013. http://www.reuters.com/article/us-highfrequency-microwave
-idUSBRE94001920130501.

Ouma, Stefan, Leigh Johnson, and Patrick Bigger. "Rethinking the Financialization of
'Nature.'" *Environment and Planning A: Economy and Space* 50, no. 3 (2018): 500–511.

Pickren, Graham. "The Factories of the Past Are Turning into the Data Centers of the
Future." *Conversation*, January 3, 2017. https://theconversation.com/the-factories
-of-the-past-are-turning-into-the-data-centers-of-the-future-70033.

Pletz, John. "Coresite Buys Downtown Site for Data Center." *Crains*, February 15, 2018.

https://www.chicagobusiness.com/article/20180215/CRED03/180219925/coresite
-buys-downtown-site-for-data-center.

Robertson, Morgan. "The Nature That Capital Can See: Science, State, and Market in
the Commodification of Ecosystem Services." *Environment and Planning D: Society and Space* 24, no. 3 (2006): 367–87.

Ryan, Tom. "Sears Replaces Retail Stores with Data Centers." *Forbes*, May 13, 2013. http://
www.forbes.com/sites/retailwire/2013/05/31/sears-replaces-retail-stores-with-data
-centers/#703dfc496b0a.

Scott, Allen J. "Beyond the Creative City: Cognitive–Cultural Capitalism and the New
Urbanism." *Regional Studies* 48, no. 4 (2014): 565–78.

Smith, Andrew. "Fast Money: The Battle against the High Frequency Traders." *Guardian*,
June 7, 2014. https://www.theguardian.com/business/2014/jun/07/inside-murky
-world-high-frequency-trading.

Sverdlik, Yevgeniy. "Digital to Convert Motorola's Former Chicago HQ to Data Center." *Data Center Knowledge*, August 10, 2016. http://www.datacenterknowledge
.com/archives/2016/08/10/digital-realty-buys-campus-in-tight-chicago-data-center
-market/.

Timms, Aaron. "Feeding the Need for Speed." *Daily Deal* 23, no. 189 (2012): 5–12.

TWO

An Emerging Satellite Ecosystem and the Changing Political Economy of Remote Sensing

Luis F. Alvarez León

For the past half century remotely sensed satellite data has become a fundamental tool available to scientists the world over—especially those in developed countries. Collected, maintained, and distributed through satellite data infrastructures underwritten primarily by national governments, such as Landsat (United States) and Satellite Pour l'Observation de la Terre, or SPOT (France), this data is now a staple of scientific work, policy-making, and public discourse. By providing robust and reliable access to repositories of Earth observations, satellite data infrastructures have been instrumental in enabling paradigm shifts in scientific knowledge about our planet, such as the detection and close monitoring of global climate change.

Yet for most of their existence, satellite data infrastructures did not change dramatically and remained under the control and management of a handful of governments and their industry partners. In recent years, however, conditions have aligned to accelerate a number of overlapping transformations in the satellite industry, with impacts on the quality, availability, accessibility, monetization, and geopolitical implications of remotely sensed data. While much well-deserved attention has been focused on the technological leap represented by small and miniaturized satellites, some of which weigh less than a kilogram, this innovation cannot be understood outside the political economic context of other transformations underway.

This chapter provides an overview of the changing political economy of satellites and remote sensing by considering the influence and interplay of three key transformations that are simultaneously expanding and upending existing satellite data infrastructures: the rise of small-satellite firms in a new digital economy, along with attendant debates about data privacy, security, and commercialization; geopolitical shifts behind the growth of satellite programs from China and India to Kenya and Nigeria; and underlying these, the ongoing rearrangements between the state and private sectors associated with economic neoliberalization.

The result of these multiple transformations is what I refer to as an emerging satellite ecosystem. This ecosystem is characterized by an expanded field of actors involved in satellite technologies and Earth observation, including start-ups and other private entities, universities, and governments from a growing number of countries. Many of these actors are drawn by a renewed drive to commercialize remote-sensing platforms and satellite data within the context of the growing digital economy. This economy has been spurred by the decades-long process of neoliberalization, which has significantly redefined the role of the state and other actors in Earth observation (as well as a wide range of other areas), creating more incentives for profit-oriented activities and expanding the scope of commodification.

Since their initial launch in the mid-twentieth century, satellite programs have been symbols of national pride, strategic military assets, and increasingly, essential infrastructure for communication and scientific research. Remote-sensing imagery collected by satellites has become a fundamental tool to examine Earth from a vantage point that promises both wide territorial scope and the collection of observations unmediated by political and social factors on the ground. However, in spite of the universalizing impetus that often colors rhetoric around satellites and remote-sensing data, these are in fact deeply enmeshed in political and social factors and thus cannot be considered unmediated in any meaningful way. For example, even though remote-sensing imagery is a key input for scientific research, the scientific community experiences differential access to it due to geographic location, institutional affiliation, resource allocation, and technological availability. This asymmetry is shaped by the decisions, policies, and capabilities of national governments in funding, collecting, and granting access to imagery, as well as the acute strategic and geopolitical importance of satellites and remote-sensing data.

Alvarez León

Even as states exercise a gatekeeping role, they have simultaneously invested large sums of resources in distributing satellite data for public use through repositories associated with Landsat and SPOT. These arrangements are complicated by ongoing changes in the funding structures of satellite programs, public-private partnerships, new start-up initiatives, changing regulatory regimes, and monetization schemes that shape access and use of satellite data. As has been the case across other categories of geospatial data, media, and technologies, the political economy of satellite programs is now experiencing a transition toward diversification, along with a highly specialized selection of industry partners, producing a new satellite ecosystem.

While it was the United States and the USSR that opened the space race in the late 1950s, other countries developed satellite capabilities in the following decades, such as France (1965), Japan (1970), China (1970), the United Kingdom (1971), other member states of the European Space Agency (1979), India (1980), and Israel (1988). Since the end of the Cold War, this configuration—dominated by a small subset of powerful states with capabilities to deploy and maintain satellite infrastructures—has begun to shift. According to data from the Union of Concerned Scientists (UCS) Satellite Database, 4,550 satellites were orbiting Earth as of September 1, 2021.[1] More than half of all these satellites are operated by the United States (2,788), Russia (167), and China (431), which reflects both the Cold War path dependency of satellite technologies and their overlap with contemporary superpower geopolitics. However, while these legacies continue to shape the satellite ecosystem, dozens of other countries and international organizations now have a presence in space.

This expansion in the geography of satellite capabilities has been matched by a diversification in the range of applications. While satellite programs started as government endeavors, today only 490 are used exclusively for this purpose, and an additional 402 are classified solely as military. There are also 146 civil satellites owned and operated mostly by universities around the world. However, these single-purpose satellites are dwarfed by the growing presence of commercial satellites, which currently number 3,207, well more than the previous three categories put together.[2] While not a full accounting of the U C S dataset—and joint-use satellites in particular—these numbers provide a sense of the growth and diversification of the satellite ecosystem.

Both the geographic expansion of satellite access and their functional diversification are the products of changing geopolitical, technological, and political economic conditions, which have accelerated the transformation of the satellite ecosystem. The rest of this chapter examines the environment resulting from such changing conditions, along with the production, distribution, and use of remotely sensed data that takes place within it. The next section examines remote sensing in the context of the rise of small satellites and the broader expansion of the digital economy.

Remote Sensing, Small Satellites, and the Digital Economy

While remote sensing has catalyzed a paradigm shift in Earth observation and analysis, it has not been able to escape the conundrum of disentangling pattern and process—that is, how things are arranged in space versus how these configurations change over time.[3] Since the launch of the first Landsat mission in 1972, satellite observations have become more frequent, and their imagery archives have drastically improved in both temporal and spatial resolution. This means not only that satellites are continuously providing sequences with ever shorter intervals that better approximate the changes in Earth's systems, but also that these sequences are made up of images that provide an increasing level of detail. Yet a real-time view of Earth continues to escape most satellite programs—a goal that has become more attainable with the contemporary transformations that are the focus of this chapter.

An example of both continual improvement and persistent limitations of satellites can be seen in Landsat 9. Launched in 2021, this satellite is in many ways superior to previous Landsat iterations in terms of the imagery that it collects, although it is limited to producing a view of Earth's entire surface every sixteen days.[4] A significant gap continues to exist between static yet detailed snapshots of any particular place and the dynamic processes that such images are only partially able to capture. This mismatch means fundamental blind spots about

Alvarez León

how events unfold on the ground (particularly those at timescales on the order of hours or days) and how to address them.

Matt Turner has summarized the issues that this problem creates in terms of the knowledge we can gain from satellite data. He highlights the tensions between environmental scientific traditions and the use of remotely sensed data in studying the Sahel, a semiarid region in the southern-central latitudes of Northern Africa: "Similar to the environmental scientific tradition in the Sahel, understandings of environmental change supplied by remotely sensed data are visually descriptive rather than processual. . . . Therefore, there is a tendency for environmental change to be described as changes in land-use or vegetative cover, with little attempt to link this to underlying causes of these changes nor to how they affect the productive capacity of the land."[5] In Turner's view, this gap created by relying on the patterns captured by satellite data should be addressed by complementing remote sensing and geographic information systems with in-depth on-the-ground analyses that can explain how processes unfold.

Yet ongoing advances in satellite technology reiterate the promise to more closely track process through improved remote-sensing technologies. The recent technological leap represented by small satellites underscores this aspiration by qualitatively shifting how remote-sensing imagery might inform a processual understanding of the world. This is to be achieved through the combination of low-cost and high-frequency imagery capture afforded by constellations of small, coordinated satellites that together can monitor large areas simultaneously and scan the entire planet at an unprecedented rate.

Will Marshall, a former NASA scientist and cofounder of Planet—a leading firm in the budding small-satellite industry—describes his company's mission by pointing to a fundamental problem with remote sensing, the "data gap" that exists between observation, reality, and action: "You can't fix what you can't see." For him a key component of the problem is the lack of appropriate tools for observation, which prevent the right course of action; in other words, the problem is "big, expensive, and slow" satellites.[6] The solution put forth by Planet, and other firms in the small-satellite industry, is a production model premised on ultracompact, relatively inexpensive satellites, which can be manufactured and launched at scale.

This Silicon Valley start-up aims to disrupt what it sees as a lack of access to Earth imagery, which Marshall says has been "the province of big governments and big companies."[7] Planet's approach, described as "Aerospace know-how meets Silicon Valley ingenuity," consists of an "agile" business model, a humanitarian orientation, and a rhetoric of openness—a combination in line with the technological utopianism that characterizes the ideology prevalent in Silicon Valley and the information technology sector at large.[8] Planet, whose motto is "insights at the speed of change," as well as the small-satellite industry more generally, has identified its niche—and the corresponding opportunities for disruption—by addressing in its "Mission 1" the very processual shortcomings of remote sensing identified earlier by Turner: "Our mission is to image the entire Earth every day, and make global change visible, accessible, and actionable."[9] Planet's Dove Constellation, which as of April 2018 consisted of 207 satellites, hovers at a stationary orbit around Earth and uses the planet's rotation to cover the entire surface every twenty-four hours, effectively acting as a "line scanner" that produces 1.5 million images covering 350 million square kilometers every day.[10] According to Marshall, Planet Labs "doesn't take a picture of anyone on the planet every day, [it] take[s] a picture of every single place on the planet every day."[11] As advertised by the company, these satellites can capture urban growth, water security, food security, deforestation, ice caps melting, floods, fires, and earthquakes on a daily basis.

The range of applications of Planet's network of small satellites is a reflection of the dual commercial and humanitarian objectives animating this company's activities. For example, one of Planet's stated objectives is to democratize access to the new global dataset collected by its satellite network, which it implicitly contrasts with existing imagery repositories.[12] However, while Planet is enabling a paradigm shift by assembling higher temporal resolution datasets and making them available to the public, its messaging elides two core issues. The rhetoric surrounding Planet's innovative model underplays how an established program like Landsat has constructed and continues to expand

Alvarez León

the longest-running repository of continuous Earth observation imagery, which is both publicly funded and already publicly accessible. Planet's message of providing a public good through "universal access" is also in tension with the private nature of this company's enterprise and its underlying profit imperative. This suggests that Planet is in similar territory as IT industry leaders such as Facebook, Google, and Amazon, which are deeply invested in the monetization of (personal and public) data collection and surveillance that characterize the contemporary digital economy.

New remote-sensing imagery and other data repositories, such as those built by Planet, should be understood as part of a growing satellite industry that includes established leaders such as SpaceX and Jeff Bezos–founded Blue Origin, as well as up-and-coming firms like U.S.-based Terran Orbital, Netherlands-based Hiber, and Australia-based Myriota. Many of these firms explicitly frame their satellite services in the context of the Internet of Things. This suggests that the satellite industry is becoming increasingly embedded in the digital economy. In particular, satellite data can be contextualized in terms of the growing supply, and rising demand, of highly detailed georeferenced data that extracts, identifies, and sorts location as well as context.[13] Read in this way, to understand the new satellite ecosystem, it is necessary to weigh the benefits brought by technological advances in Earth observation against the sociopolitical and economic consequences of an increasingly privately held spatial repository of "every single place on the planet every day." A step toward making that assessment is to examine the disruptions created by small-satellite technologies in relation to contemporary geopolitical shifts.

Geopolitical Shifts and Emerging Satellite Programs

In the context of a digital economy with a high demand for geospatial data, the production of remotely sensed imagery calls for examination of its geopolitical dimensions. As it continues to feed a growing market for geospatial data, remotely sensed imagery is intrinsically tied to large-scale geopolitical shifts and articulations of power, such as the ballooning surveillance apparatus fueled by big data.[14] In the current satellite ecosystem, the entanglement of privatization, scientific value, military applications, and surveillance capabilities shapes the role of remotely sensed imagery as an informational product in the geospatial digital economy. Yet the complex implications of this entanglement are often conspicuously absent from the commercial-humanitarian rhetoric that delivers this data into the marketplace. In this section I sketch out some of the geopolitical dimensions of the new satellite ecosystem and emphasize how the changing remote-sensing landscape is both driven by and intersects with politics at various scales.

The collection of remote-sensing data entails capturing from afar the features of any given surface or territory. While in scientific analysis the political dimensions of this activity are generally not the most salient, it is essential to reckon with the fact that data acquisition and analysis, in and of itself, represents a political intervention. This is because to produce a satellite image about a particular location, it is necessary to have access to it (even if it is at a distance), which comes with a degree of power that is augmented by the collection of information. It is no coincidence that since their inception, satellites have been used by states for military and intelligence purposes. This applies not only to those satellite programs that have been under the direct control of governments but also to private corporations with agreements to provide imagery to military agencies, such as Planet with the U.S. National Reconnaissance Office.[15] Thus while remote sensing entails both the observation of features and events on the ground and analysis to inform decision-making, the very act of satellite observation invariably carries with it a "politics of witnessing" through which various actors attempt to regulate the meaning of the imagery and events themselves.[16]

These inescapable politics present in all remote sensing can be seen, for instance, in the development of a technique known as at-many-stations hydraulic geometry (AMHG). This is a way to collect river flow rate estimates solely from remote-sensing imagery. Such a technique is a significant innovation because river flow rate estimates have traditionally required physical access to rivers for the placement of measurement gauges at various cross sections. This has led to difficulties in measurement and incomplete datasets that are further curtailed by the access restrictions placed on rivers by various governments around the world because of the strategic nature of river flow data—particularly in cross-boundary cases.

Through their discussion of AMHG and its deployment on the international river basins of the Ganges-Brahmaputra and the Mekong, Colin Gleason and Ali Hamdan demonstrate that the collection of remotely sensed imagery can have politically charged meaning and effectively constitutes a political intervention at various scales, while also representing unambiguous scientific value.[17] This particular example illustrates how satellites can make accessible what is deliberately made inaccessible by other means due to the territorial control over resources exercised by specific actors, especially states. In this case, since many governments around the world consider river flow data to be a state secret, the authors limit their analysis to data from river gauges that are publicly available. This decision is presumably informed by the possibility that publishing river flow estimates would directly affect the international relations between neighboring states, particularly between those who have disputes over river water use (such as India with Bangladesh, China with Southeast Asian states downstream of the Mekong, and Ethiopia with Egypt and Sudan). The authors explicitly articulate their political decision to withhold calculation and distribution of specific data, while simultaneously releasing the methods to conduct such calculations: "By not calculating flow rates for contentious or ungauged portions of basins, we hope to demonstrate the efficacy and utility of AMHG flow rate retrievals and characterise the policy environment in each case without imposing AMHG upon local actors who will, ultimately, determine its usage and success."[18]

The strategic and economic value of remotely sensed imagery and data derived from it, such as river flow estimates, creates incentives for new satellite infrastructures that circumvent the financial and technological arrangements that previously defined the satellite industry and traditionally reflected the core-periphery divisions of the global political economy. While new market opportunities in the private sector are contributing to the expansion of availability and cost viability of new satellite technologies, there is a growing roster of countries that are taking advantage of these conditions to develop satellite programs and partake in the potential strategic and monetary benefits.

In the past decades rising powers such as India and China developed full-fledged space programs, while many other countries invested in having a presence in outer space through satellites. The ninety-five countries and country alliances with satellite capabilities include North-North alliances, North-South alliances, South-South alliances, regional organizations, and countries in every continent, from Mexico and Brazil to Nigeria, South Africa, Indonesia, Laos, and Kazakhstan. The distribution of these satellites by country or country alliance is depicted in figure 4. The full list of countries and satellites can be found in table 1 in the appendix at the end of this chapter.

This new space race is particularly significant for developing countries, which have to contend with unprecedented challenges of managing costly and complex sociotechnical organizations while balancing urgent developmental outcomes and the potential benefits that might be derived from homegrown satellite programs.[19] An illustrative example is the impact of new satellite programs in African geopolitics, since a number of countries across the continent are devoting important resources to developing a range of space capabilities, from Earth observation to proposed manned missions, the latter of which has been announced by Nigeria for the year 2030.[20] As relative newcomers to the space race, countries including South Africa, Algeria, Ethiopia, Morocco, and Nigeria are taking advantage of the expanding field of private partnerships and international alliances that allow them to translate their geopolitical aims and technological ambitions into orbit.

These collaborative arrangements are within the arenas of international or scientific cooperation, as well as the satellite technology global marketplace. But they are also typically shaped by both colonial legacies and newer articulations of regional power, to which they can actively contribute. For example, Algeria's satellite Alcomsat-1 is the result of a long-term partnership with China, while Morocco's Mohammed VI-A and VI-B satellites were developed through contracts with French companies, a partnership that should be understood in the context of the two countries' colonial past.[21] Egypt's EgyptSat 2, on the other hand, was manufactured by a Russian company. It is actively used to moni-

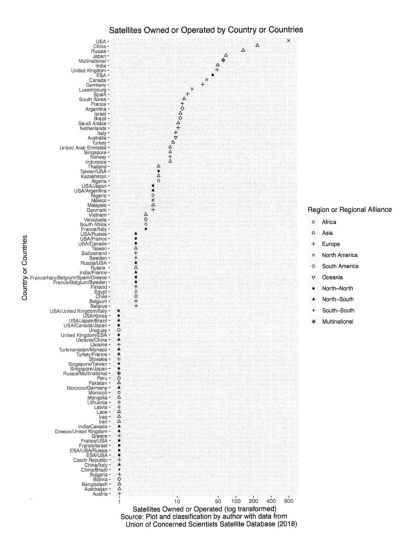

4. Number of satellites owned or operated by country or group of countries. Plot and classification by author with data from Union of Concerned Scientists Satellite Database (2018).

tor Ethiopia's water consumption along the Blue Nile and its construction of the Renaissance Dam, a focal point of tension because of the potential water shortage the dam may cause in lower Nile states such as Sudan and Egypt.[22]

The shifting geopolitics of these newly satellite-capable states have prompted the African Union to convene meetings toward the creation of an African Space Policy to regulate the emerging satellite and space programs, ensure cooperation, and minimize the potential for conflict that can arise from them.[23] While this framework is in its early stages, it highlights how satellite capabilities can influence the intra- and extra-regional arrangements of the continent by endowing national governments with, among other assets, domestic and international surveillance capabilities that upset the balance of power along both political and economic planes. A closer look at any of these newly developed space and satellite programs reveals how the emerging satellite ecosystem in Africa is jointly produced by the lasting influence of colonialism and by new technological and political economic conditions. This has important implications not only for the geopolitical dimensions of satellite data but also for how it intersects with economic incentives, scientific endeavors, and legacies of power and domination across various scales.

One example of these configurations is Kenya's first home-designed miniature satellite, the 1st Kenyan University NanoSatellite-Precursor Flight (1KUNS-PF). This is a cube satellite made of miniature cubic units, each of which weighs little more than a kilogram. The satellite went into orbit in May 2018 and was the result of a coordinated effort among a number of actors that exemplify the emerging satellite ecosystem: the Kenyan scientific community, a U.S.-based private company, and various international scientific cooperation partnerships. The satellite was designed by engineering faculty and students at the University of Nairobi, with technical assistance from personnel at the Sapienza University in Rome. The project, sponsored by the United Nations Office for Outer Space Affairs and the Japan Aerospace Exploration Agency, was conducted within the framework of collaboration between the National Space Secretariat of Kenya and the Italian Space Agency, which provided joint support.[24] The satellite was launched into space by a SpaceX rocket on a resupply mission to the International Space Station. Subsequently, it was launched into orbit from the Japanese module of the space station.[25] While this event is part of an important turning point for the opening of the satellite field to developing countries, it is thoroughly informed by previous historical episodes, since the conditions they created continue to influence the rapid changes shaping the political economy and technology of the emerging satellite ecosystem.

According to John N. Kimani, the lead scientist at Kenya's National Space Secretariat, the IKUNS-PF project was conducted within the "renewal of San Marco agreement between Kenya and Italy."[26] This agreement is significant because, among other reasons, it represents a longer arc of collaboration between these two countries that has had an important impact on satellite development. It is also an expression of colonial injustices and core-periphery asymmetries that pervade science and technology partnerships. The San Marco agreement was signed in 1962 between Italy and the British administration of Kenya with the purpose of allowing Italian scientists to build a launching site and support infrastructure on Kenyan territory.

Kenya became an independent state in 1963, and the following year the new government signed a treaty with Italy that addressed the offshore construction of a launching platform in the Indian Ocean, near the southeastern Malindi area, along with a base camp and logistical facilities in the mainland. The establishment of the San Marco platform had two key scientific and operational objectives: achieving the first equatorial launch to collect atmospheric data along the equator and establishing a base along that latitude that could support future launches.[27] While initially the Italians built the platform just outside the three-mile fringe of Kenyan territorial waters, the Territorial Waters Act of 1972 extended this limit to twelve nautical miles. This placed the site of the platform squarely within Kenya's territory.[28]

The San Marco agreement was renewed several times in the subsequent decades, allowing for the launch of dozens of rockets and satellites. Among these were the Uhuru in 1970, the first X-ray-capable satellite ever launched and the first U.S. satellite launched by another nation.[29] Originally named Small Astronomy Satellite 1 (SAS-1), this was in effect an American satellite launched by Italian personnel through a collaboration between NASA and the Italian Space Agency. However, in recognition of the host country, the satellite was later renamed Uhuru, the Swahili word for freedom, which was an attempt to link it more explicitly to Kenya's history and independence. While Uhuru's historic launch was a success and paved the way for the launch of many more American, British, and Italian satellites, the glaring omission in the history of the San Marco agreement has been the lack of active participation by Kenyans in the operation and decision-making of this program.

This omission prompted heated debate in the Kenyan National Assembly regarding the conditions of the initial 1962 and 1964 agreements, the benefits derived from them, and the exclusion of Kenyans from participating in the program. In particular, members of the Kenyan Parliament such as Wilber Ottichilo, a space scientist who led an investigation into the San Marco facilities, pointed out that several countries had been making commercial use of the launch facility but the Kenyan government was neither notified nor compensated for the economic or scientific benefits this generated. In addition, there were unresolved questions regarding data collection and ownership, as well as allegations of a lack of transparency on the part of Kenya's Department of Defense, which had handled the operations.[30]

After these questions were raised, a joint committee was formed in the Kenyan Parliament to explain "the status of the San Marco Space Application Center in Malindi," also known as the Luigi Broglio Space Center. The committee's findings suggested that the San Marco treaty should be reexamined to address the uneven distribution of benefits and data, along with underrepresentation of the Kenyan personnel center's workforce.[31] In the debate that ensued at the National Assembly in 2014, Aden Bare Duale, a member of Parliament and then majority leader, addressed the core asymmetry at the heart of the San Marco agreement and the role of the Italian government:

> I want to confess, go on record and say that, last year, the Italian Ambassador visited me. In one of our joint discussions, he told me that if we do not sign the agreement, then our bilateral relationship will be affected.

We do not want a penny from the Italian Government. What we want is that resource. They should give it back and pay us reparations. Hon. Temporary Deputy Speaker, technology transfer increases access—we should even go further and take over the site. Why should we cost-share with them? Do we have a similar San Marco Space Application Centre in Milan or Rome?[32]

The inequities underlying the San Marco agreement constitute an issue that crosses party lines, eliciting agreement across the political spectrum. Another member of Parliament, Nicholas Gumbo, echoed this sentiment by summarizing the economic implications of access and agency over satellite data:

You ask yourself why the Government of Kenya has not insisted on getting a share of the billions of shillings that, that Centre generates from third party contracts. You can go a step further and ask yourself why our successive Governments did not find it necessary to use the massive amounts of data that is generated from third party contracts for our own technological advancement. These are questions that we kept on asking when we went to San Marco. Hon. Temporary Deputy Speaker, it is a known fact that, that Centre generates billions of shillings. We spent the whole day there asking how much does the Centre generates. In fact, the impression the Italians running the Centre were trying to give us was that the Centre was a non-profit Centre. It is laughable! It is ridiculous and unacceptable. This Centre, for the information of hon. Members and those who may not know, has been used to launch rockets from the Kenyan landscape.[33]

These statements and the ongoing debates in the Kenyan National Assembly regarding the San Marco agreement are informed by the geopolitical asymmetries in the development and use of satellites. The distribution of benefits arising from such asymmetries is brought to the fore by the rising monetary value of remote-sensing and other types of data in the digital economy. In this context the successful development and launch of IKUNS-PF represents a step forward for a Kenyan space program with a more substantive role played by local actors. Together these contrasting dimensions shape the long arc of space technology carried out in Kenya, if not predominantly *by* Kenyans. The same unequal core-periphery relationship that gave way to the San Marco agreement in the first place is now serving as a basis for cooperation and development agreements, such as those that led to IKUNS-PF. Yet it is not clear what to make of the benefits that such institutional structure has produced over the course of half a century, and it is still uncertain how these might be distributed moving forward. However, the Kenyan Parliament approved an "equal status" deal in July 2019, which, if implemented, would make the Italian government pay their Kenyan counterpart fees for land rent for the base (Sh 25 million, or around US$250,000, with US$50,000 increments every five years) and annual authorization from each third-party use (Sh 5 million, or around US$50,000).[34]

Even though the satellite ecosystem is expanding and more countries are acquiring Earth observation capabilities, the dynamics that have characterized the Kenyan case are illustrative of the core-periphery geopolitics that have shaped the production of and access to satellite technologies since their inception. It is for these reasons that we should evaluate satellite data infrastructures in their broader historical and geographic context. In doing so, we must keep asking how satellite imagery is produced within particular political economic arrangements, who benefits from it, and how the imagery stands to change the balance of power in these arrangements.

Conclusions

The second organizational mission of Planet, currently underway, is to enhance remote-sensing imagery with artificial intelligence, producing a searchable database of Earth "in the same way Google indexed the internet."[35] This project is in a similar vein to the efforts of another technological giant, Facebook, to produce highly detailed population maps using satellite imagery combined with artificial intelligence.[36] These datasets can help keep closer track of world trade, conservation efforts, disaster relief, development, and many other important issues. However, these technologies can readily be deployed for surveillance, repression, harassment, and exploitation of people and the environment, as has been repeatedly shown to be the case with large-scale information networks and digital platforms. Thus remote-sensing imagery is becoming wholly integrated into the digital information economy, while the emerging satellite ecosystem is being shaped by the same tensions between public and private goods, and among privacy, profit, security, and surveillance, that affect other types of commodified data.

And while remote sensing is coming ever closer to realizing the promise of providing a real-time record of Earth, the conundrums that characterize knowledge derived from satellite imagery remain. In particular, as Turner has pointed out, no matter how closely satellites may track a phenomenon, we must continue to pay attention to the differences between pattern from above and process on the ground, retaining a sense of critical analysis to interrogate, triangulate, or complement the claims made through the use of satellite data, no matter its spatial and temporal resolution.[37] This is particularly important given that it might be plausibly claimed that with their high temporal and spatial resolution, satellites can show the world "as-is."

To avoid falling into the trap of thinking of satellite data as unmediated, or free from the influence of extraneous factors, Paul Robbins and Tara Maddock remind us to engage in a process of denaturalization.[38] This is to highlight and explicitly discuss the artificial and artifactual aspects of satellites and the data they produce. Despite the increasingly sophisticated technologies for capturing, processing, indexing, and analyzing satellite data, it is still the product of human decisions, technical arrangements, and social conditions, all of which are influenced by individual and collective choices and geopolitical circumstances.

In fact, these circumstances are at the foundation of the emerging satellite ecosystem and are in many ways influenced by historical patterns of power and knowledge, such as colonialism, Cold War geopolitics, and uneven development. In light of these factors, and in the context of a growing interest in amassing, analyzing, and commercializing remote-sensing data, it remains urgent to ask—and demand answers to—questions such as who is in control of satellite information, what it can be used for, and who can benefit from it. As the cases presented in this chapter illustrate, we should also be concerned about who is vulnerable to and who can be affected by the very acts of producing and distributing satellite imagery, as well as its potential misuses, asymmetric appropriation, exploitation, and control.

Appendix

TABLE 1. Satellite ownership or operation

Country or partnership	Satellites owned or operated
Algeria	5
Argentina	30
Australia	14
Austria	1
Azerbaijan	2
Bangladesh	1
Belarus	2
Belgium	2
Bolivia	1
Brazil	13
Bulgaria	2
Canada	53
Chile	1
China	462
China/Brazil	2
China/France	2
China/Italy	1
Colombia	1
Czech Republic	3
Denmark	4
Ecuador	1
Egypt	4
ESA	60
ESA/USA	1
ESA/USA/Russia	1
Estonia	1
Ethiopia	2

Finland	15
France	17
France/Belgium/Sweden	2
France/Israel	1
France/Italy	3
France/Italy/Belgium/Spain/Greece	2
France/USA	1
Germany	44
Greece	2
Greece/United Kingdom	1
Hungary	1
India	58
India/Canada	1
India/France	2
Indonesia	8
Iran	1
Iraq	1
Israel	18
Italy	14
Japan	84
Japan/Singapore	2
Jordan	1
Kazakhstan	6
Kuwait	1
Laos	1
Lithuania	2
Luxembourg	40
Malaysia	4
Mauritius	1
Mexico	7
Morocco	2
Morocco/Germany	1
Multinational	62
Nepal	1
Netherlands	15

New Zealand	1
Nigeria	3
Norway	8
Pakistan	3
Paraguay	1
Peru	1
Poland	2
Qatar	1
Russia	165
Russia/USA	2
Saudi Arabia	13
Singapore	8
Singapore/Taiwan	1
Slovenia	2
South Africa	3
South Korea	18
Spain	22
Sri Lanka	1
Sudan	1
Sweden	2
Switzerland	13
Taiwan	1
Taiwan/USA	11
Thailand	7
Tunisia	1
Turkey	9
Turkmenistan/Monaco	1
Ukraine	1
United Arab Emirates	13
United Kingdom	345
United Kingdom/ESA	1
United Kingdom/Netherlands	1
USA	2,769
USA/Argentina	4
USA/Canada	2

USA/Canada/Japan	1
USA/France	1
USA/Germany	2
USA/Japan	5
USA/Japan/Brazil	1
USA/Mexico	1
USA/Sweden	1
USA/United Kingdom/Italy	1
Venezuela	2
Vietnam	4
Total	4,550

Source: Created by author with data from Union of Concerned Scientists Satellite Database (2021).

Notes

1. UCS, "UCS Satellite Database."
2. UCS, "UCS Satellite Database." A single satellite can have multiple uses, defined by the affiliations of its primary users, and there are many joint-use satellites that are not counted in this classification.
3. O'Sullivan, "Pattern, Process and Scale."
4. Smith, "Landsat 9." In conjunction with Landsat 8, the two satellites revisit any point on Earth every eight days.
5. Turner, "Methodological Reflections," 261–62.
6. Marshall, "Tiny Satellites." Planet was known as Planet Labs until 2016.
7. Marshall, "Planet Labs' Will Marshall."
8. Planet, "Our Approach"; Turner, "How Digital Technology Found Utopian Ideology."
9. Planet, "Planet Imagery and Archive"; Planet, "Planet at a Glance."
10. Marshall, "Mission to Create."
11. Marshall, "Tiny Satellites."
12. Marshall, "Tiny Satellites."
13. Alvarez León, "Property Regimes"; Leszczynski, "Situating the Geoweb."
14. Crampton, "Collect It All."
15. Marcus, "National Reconnaissance Office."
16. Parks, "Satellite Views of Srebrenica."
17. Gleason and Hamdan, "Crossing the (Watershed) Divide."
18. Gleason and Hamdan, "Crossing the (Watershed) Divide," 12.
19. Wood and Weigel, "Building Technological Capability"; Wood and Weigel, "Architectures."
20. Monks, "Nigeria Plans to Send an Astronaut."

21. Russell, "Algeria Joins the Space Club"; Clark, "Moroccan Spy Satellite."

22. Egyptian Streets, "Egypt's New Satellite."

23. African Union, *African Space Policy*.

24. Mbuthia and Ouma, *IKUNS-PF*.

25. Dahir, "Kenya Heads into Space."

26. Mbuthia and Ouma, *IKUNS-PF*, 3.

27. Nesbitt, *History*.

28. Owaahh, "Space Center Kenya Doesn't Own"; Territorial Waters Act of 16 May 1972, revised in 1977, Parliament of Kenya, https://www.un.org/Depts/los /LEGISLATIONANDTREATIES/STATEFILES/KEN.htm.

29. Nesbitt, *History*, 1.

30. Kwayera, "Kenya-Italy Space Agreement."

31. COFEK, "How the Italian Government Has Duped Kenya."

32. Kenyan National Assembly, *Official Report*.

33. Kenyan National Assembly, *Official Report*.

34. Ibeh, "Kenya and Italy."

35. Baylor, "Planet Labs."

36. Slatt, "Facebook Is Using AI."

37. Turner, "Critical Reflections."

38. Robbins and Maddock, "Interrogating Land Cover Categories."

Bibliography

African Union. *African Space Policy: Towards Social, Political, and Economic Integration*. HRST/STC-EST/Exp./15 (II):15. Cairo: African Union, 2017. https://au.int/sites /default/files/newsevents/workingdocuments/33178-wd-african_space_policy_-_st20444_e_original.pdf.

Alvarez León, Luis F. "Property Regimes and the Commodification of Geographic Information: An Examination of Google Street View." *Big Data & Society* 3, no. 2 (2016): 1–13.

Baylor, Michael. "Planet Labs Targets a Search Engine of the World." *NASASpaceflight. com*, January 29, 2018. https://www.nasaspaceflight.com/2018/01/planet-labs-targets -search-engine-world/.

Clark, Stephen. "Moroccan Spy Satellite Launched Aboard Vega Rocket." *Spaceflight Now*, November 21, 2018. https://spaceflightnow.com/2018/11/21/moroccan-spy -satellite-launched-aboard-vega-rocket/.

COFEK (Consumers Federation of Kenya). "How the Italian Government Has Duped Kenya, Continues to Make a Kill from the San Marco Space Application Centre in Malindi as Locals Get a Raw Deal." May 26, 2014. https://www.cofek.africa/2014 /05/how-the-italian-government-has-duped-kenya-continues-to-make-a-kill-from -the-san-marco-space-application-centre-in-malindi-as-locals-get-a-raw-deal/.

Crampton, Jeremy W. "Collect It All: National Security, Big Data and Governance." *Geo-Journal* 80 (2014): 519–31.

Dahir, Abdi Latif. "Kenya Heads into Space with the Launch of Its First Home-Designed

Cube Satellite." *Quartz Africa*, May 11, 2018. https://qz.com/africa/1275698/kenya
-to-launch-first-satellite-into-space/.

Egyptian Streets. "Egypt's New Satellite Spies on Australia." *Egyptian Streets* (blog),
May 12, 2014. https://egyptianstreets.com/2014/05/12/egypts-new-satellite-spies
-on-australia/.

Gleason, Colin J., and Ali N. Hamdan. "Crossing the (Watershed) Divide: Satellite Data
and the Changing Politics of International River Basins." *Geographical Journal* 183,
no. 1 (March 2017): 2–15.

Ibeh, Joseph. "Kenya and Italy Close In on Signing Ownership Deal in Respect of Luigi
Broglio Space Centre." *Space in Africa* (blog), July 19, 2019. https://africanews.space
/kenya-and-italy-signing-ownership-deal-luigi-broglio-space-centre/.

Kenyan National Assembly. *Official Report*. Nairobi, Kenya, June 11, 2014. http://info
.mzalendo.com/hansard/sitting/national_assembly/2014-06-11-14-30-00.

Kwayera, Juma. "Kenya-Italy Space Agreement on Hold after State Questions Viability."
Standard, August 12, 2012. https://www.standardmedia.co.ke/article/2000063889
/kenya-italy-space-agreement-on-hold-after-state-questions-viability.

Leszczynski, Agnieszka. "Situating the Geoweb in Political Economy." *Progress in Human
Geography* 36, no. 1 (February 2012): 72–89.

Marcus, Jen. "National Reconnaissance Office Signs Contract with Planet Federal." *Planet
Pulse*, June 3, 2019. https://www.planet.com/pulse/national-reconnaissance-office
-signs-contract-with-planet-federal/.

Marshall, Will. "The Mission to Create a Searchable Database of Earth's Surface." Filmed
April 2018 in Vancouver BC. TED video, 6:13. https://www.youtube.com/watch?v
=IQkj4CF_ha4.

———. "Planet Labs' Will Marshall on Launching 150 Earth-Imaging Satellites into
Orbit." Filmed March 2016. This Week in Startups video, 4:27. https://www.youtube
.com/watch?v=Ztu2M8J_MIE.

———. "Tiny Satellites That Photograph the Entire Planet, Every Day." Filmed March
2014 in Vancouver BC. TED video, 8:05. https://www.youtube.com/watch?time
_continue=475&v=UHkEbemburs.

Mbuthia, J. Mwangi, and Heywood Ouma. *1KUNS-PF: 1st Kenyan University NanoSatellite-
Precursor Flight*. University of Nairobi, March 31, 2016. https://vdocuments.net/1kuns
-pf-kenyan-university-nanosatellite-supplementary-notes-39-310-references.html.

Monks, Kieron. "Nigeria Plans to Send an Astronaut to Space by 2030." CNN, April 6,
2016. https://www.cnn.com/2016/04/06/africa/nigeria-nasrda-space-astronaut
/index.html.

Nesbitt, H. N. *History of the Italian San Marco Equatorial Mobile Range*. Washington DC:
NASA, 1971. https://ntrs.nasa.gov/archive/nasa/casi.ntrs.nasa.gov/19720007316.pdf.

O'Sullivan, David. "Pattern, Process and Scale." In *Spatial Simulation*, edited by David
O'Sullivan and George L. W. Perry, 29–56. Chichester, UK: John Wiley & Sons, 2013.

Owaahh. "The Space Center Kenya Doesn't Own." *Owaahh* (blog), March 31, 2016.
https://owaahh.com/space-center-kenya-doesnt/.

Parks, Lisa. "Satellite Views of Srebrenica: Tele-Visuality and the Politics of Witnessing." *Social Identities* 7, no. 4 (December 2001): 585–611.

Planet. "Our Approach." Accessed December 3, 2021. https://planet.com/company /approach/.

———. "Planet at a Glance: See Change. Change the World." August 11, 2019. https:// learn.planet.com/rs/997-CHH-265/images/Planet%20Overview%20Datasheet.pdf.

———. "Planet Imagery and Archive." Accessed December 3, 2021. https://www.planet .com/products/planet-imagery/.

Robbins, Paul, and Tara Maddock. "Interrogating Land Cover Categories: Metaphor and Method in Remote Sensing." *Cartography and Geographic Information Science* 27, no. 4 (January 2000): 295–309.

Russell, Kendall. "Algeria Joins the Space Club with New Satellite in Orbit." *Via Satellite*, December 11, 2017. https://www.satellitetoday.com/government-military/2017/12 /11/algeria-joins-space-club-new-satellite-orbit/.

Slatt, Nick. "Facebook Is Using AI to Make Detailed Maps of Where People Live." *Verge*, February 22, 2016. http://www.theverge.com/2016/2/22/11075456/facebook -population-density-maps-internet-org.

Smith, Joseph M. "Landsat 9 to Provide a Wealth of Data to the Longest Continuous Global Record of Earth Imagery." NASA Earthdata, December 16, 2021. https:// earthdata.nasa.gov/learn/articles/landsat-9-data.

Turner, Fred. "How Digital Technology Found Utopian Ideology: Lessons from the First Hackers' Conference." In *Critical Cyberculture Studies: Current Terrain Future Directions*, edited by David Silver and Adrienne Massanari, 257–69. New York: New York University Press, 2006.

Turner, Matthew D. "Critical Reflections on the Use of Remote Sensing and GIS Technologies in Human Ecological Research." *Human Ecology* 31 (2003): 177–82.

———. "Methodological Reflections on the Use of Remote Sensing and Geographic Information Science in Human Ecological Research." *Human Ecology* 31, no. 2 (June 2003): 255–79.

UCS (Union of Concerned Scientists). "UCS Satellite Database." Updated September 1, 2021. https://www.ucsusa.org/nuclear-weapons/space-weapons/satellite-database.

Wood, Danielle, and Annalisa Weigel. "Architectures of Small Satellite Programs in Developing Countries." *Acta Astronautica* 97 (April 2014): 109–21.

———. "Building Technological Capability within Satellite Programs in Developing Countries." *Acta Astronautica* 69, no. 11 (December 2011): 1110–22.

....

THREE

Smart Earth

ENVIRONMENTAL GOVERNANCE IN A WIRED WORLD

Karen Bakker and Max Ritts

W hat are the emergent possibilities for environmental governance in an increasingly wired world? Over the past two decades researchers have been creating innovations in environmental monitoring technologies that combine information and communication technologies with conventional monitoring technologies (e.g., remote sensing), and environmental sensor networks (E S N s). These digital technologies, which enable a new modality of governance that we call Smart Earth, have proliferated because of the rapid decrease in the cost of cloud-based computing and innovations in machine-to-machine infrastructure. Smart Earth technologies enable terabytes of environmental data to be derived from terrestrial, aquatic, and aerial sources. They deploy new methods of data gathering, such as E S N s, earth imaging, cell phone–based interfaces, drones, unmanned aerial vehicles, wearables, and biotelemetry, as well as new methods of data analysis, including artificial intelligence, computer vision, machine learning, and sensor data analytics.[1] Collectively, these developments have dramatically increased scientists' ability to infer changes in abiotic conditions as well as biotic communities. Newly at their disposal, an assemblage of bioacoustics technologies enables scientists to listen to the world with unprecedented spatiotemporal resolution, continuous data collection, and automated data analysis at relatively low cost.

This chapter explores some of the ways that digital technologies disrupt existing forms of environmental governance. The key forces that shape this trajectory include the unprecedented volume, integration, accessibility, and timeliness of environmental data; the proliferation of techniques for environmental sensing; the consequent potential for time-space compression of decision-making (which in turn enables automated real-time regulation and new prediction capabilities); the proliferation of new environmental governance actors; and potentially a much higher degree of transparency in environmental decision-making. Together these innovations in environmental data infrastructures create the conditions for significant transformations in environmental governance.

In making this claim, we mobilize the concept of environmental governance from an analytical, as opposed to normative, perspective, defining governance as the mode of conduct of specific institutions or organizations. We focus our analysis on the social actors and institutions, including laws, rules, norms, customs, as well as the data-gathering and decision-making processes, engaged in environmental decision-making. Our analysis draws from studies that show how decision-making power has been partially redistributed from state to nonstate actors (for example, through the emergence of nonstate market-driven governance systems) and how the rescaling of governance has occurred both above and below the nation-state.[2]

The chapter begins with a brief overview of digital technologies used in contemporary environmental governance.[3] We then discuss key issues and critiques relevant to environmental governance debates. We conclude with suggestions for future directions for environmental governance research in an increasingly digital and digitized world.

Digital Environmental Technologies: An Overview

Digital technologies used in environmental governance often combine well-established approaches, such as remote sensing and long-term ecological monitoring, with newer technologies, such as animal biotelemetry and bacteria-based biosensors, as well as modalities of data collection such as drones, Google Earth, and citizen sensing (table 2, fig. 5). The most frequent targets of these applications are natural resources (such as forests), species prioritized for conservation (such as marine mammals), earth "boundary conditions" and "ecosystem services" (such as fresh water and atmospheric carbon), and environmental security (for example, natural disasters such as floods and fires). These technologies are being created by a diverse group of actors, including large tech companies such as Google, IBM, and Microsoft; smaller tech start-ups; militaries; and partnerships between conservation nongovernmental organizations (NGOs) and industries. Within academia, new disciplines combining digital technologies, ecology, and environmental science have emerged (e.g., computational sustainability), as well as new interdisciplinary journals (e.g., *Ecological Informatics*, *Journal of Digital Earth*).[4]

TABLE 2. Examples of Smart Earth technologies

Type	Example	Description
Wearables	Flow	Flow is a smart, wearable air pollution tracker using sensors to monitor the user's real-time exposure to air pollution and an app to help users find cleaner air.
Animal biotelemetry	Save the Elephants: Geo-fencing	Geo-fencing, a tracking collar e-programmed with Global Positioning System (GPS) positions, promises real-time information on an elephant's location, potentiating new regulatory actions if an elephant moves into areas susceptible to poaching.
Plant biotelemetry (cyberplants)	PLEASED	Plants Employed as Sensing Devices (PLEASED) embeds sensors into plants to measure environmental parameters, which can be used to monitor fires and avalanches.
Insect biotelemetry	Bees with Backpacks	Small RFID "backpack" sensors on honeybees' location and movement promise insight into the decline of regional bee populations.
Mobile apps	giveO2	The giveO2 app tracks users' means of transportation, identifies their carbon footprint using GPS, and gives them the opportunity to purchase carbon credits to offset their carbon output.
Fixed sensors	Instant Detect	Instant Detect is a network of multiple fixed camera traps using a central satellite node to send photographs and data on protected animals.

Mobile sensors	Argo	Argo consists of 3,800 free-drifting ocean floats with mobile sensors that remotely transmit data on salinity and temperature from the upper two thousand meters of the ocean.
Sensor web	Ocean Observatories Initiative (OOI)	Dubbed "Fitbit for the oceans," OOI's cyber-infrastructure explores remote sections of the oceans with sensors and autonomous drones, generating over two hundred data types.
Remote sensing	GasFinder3	GasFinder3 uses laser-based remote sensing to continuously monitor gas concentrations and has been applied around oil and gas sites in the Arctic region.
Virtual reality	Conservation in Virtual Reality	A 360-degree virtual reality film "immerses" viewers in global environmental conservation efforts, including the Amazon rainforest and Indonesian reefs.
Artificial intelligence (AI)	Green Horizon	A cognitive computing system uses machine learning and real-time air quality data to analyze and create visual maps displaying the source and dispersion of pollutants across Beijing.

Source: Created by authors.

Although Smart Earth innovation has concentrated on terrestrial processes, ocean environments have also been a major focus. Across a range of ecosystems and disciplines, scientists and engineers have created new devices to assess changing oceanographic conditions, including an automated network dubbed "Fitbit for the oceans." This network incorporates cable and sensor technologies to measure geological, physical, chemical, and biological variables in the ocean and on the seafloor. It is emblematic of the ways in which Smart Earth technologies enable comprehensive data acquisition infrastructures to monitor dynamic environmental conditions, risks, and geohazards.[5]

Another paradigmatic example of Smart Oceans technologies is

EarthNC's Shark Net, which displays the near real-time locations of great white sharks and marine mammals. Here, data is gathered from a subsea network of mobile robots, moored listening stations, and animal tags. Like many tracking efforts, Shark Net emerged from collaboration among scientists interested in monitoring animals across large spatial ranges. Related citizen sensing projects have geotagged fish, including important commercial species, whose movements can likewise be detected by networks of underwater cameras.[6]

Coordinating many of these innovations is a proliferation of environmental data infrastructures that span a variety of scales and ecosystem types and seek to integrate streams of data for a range of user groups.[7] These infrastructures are often designed as focal points for certain types of environmental information, such as the soil data center at the European Commission's Joint Research Centre or the water data center at the European Environment Agency. Others are more broad-based, such as the Automatic Remote Geomagnetic Observatory System (ARGOS), which tracks marine and terrestrial species across a range of ecosystem types.

This shift toward the digitalization of all manner of environmental knowledge circulation and management, whether through the retrofitting of existing physical infrastructures (e.g., lighting, water, sanitation, public transport) or the outlay of new kinds (e.g., smartphones, fiber-optic cables, and cell-phone towers), has been quite rapid in many contexts. DataOne, which is coordinated (or "steered," to invoke governance discourse) by a network of public and private partners, including the U.S. National Science Foundation, Microsoft, and the University of California, is revealing of the "promissory" logics associated with emergent environmental data infrastructures.[8] It operates a federation of distinct data repositories and proposes to render from across disciplines, scales, and national interests the "enhanced search and discovery of Earth and environmental data." But if georeferenced, open, and accessible environmental data is now more abundant, its function remains inextricably linked to the social and material orders of the large institutions and technology firms that assemble and operate large data infrastructures. The reformatting of environmental subjects into data producers that (knowingly or unwittingly) generate millions of data points has been shown to enable "data grabs," allowing technology firms to understand specific markets and manipulate individual consumer needs.[9] Nevertheless, and through the provision of data platforms, data infrastructures can in some cases enable citizens to exchange opinions, preferences, and consumption behaviors regarding a range of environmental issues.

Environmental Governance in a Smart Earth World: Issues and Critiques

To date the majority of research on Smart Earth topics has focused on the potential implications for conservation and waste reduction, pollution mitigation, mapping environmental degradation, geosecurity, and disaster management.[10] Only a handful of scholars have engaged with the sociotechnical governance dimensions of Smart Earth.[11] This chapter seeks to address that gap by offering insights into the key implications of Smart Earth for environmental governance. In this section we explore four themes: exponential data availability, the meaning and implications of open data, the changing nature of prediction, and real-time analysis and regulation.

Exponential Data

In the past state agencies and NGOs cited the scarcity of data as a major limiting factor for effective environmental governance. In the future this constraint will be reversed: managing abundant rather than scarce data is increasingly the critical challenge. The exponential growth in environmental data generated by digital technologies is illustrated by the collaborative efforts of IBM and research scientists on Lake George in the northeastern United States.[12] This study convened biologists, environmental scientists, engineers, physicists, computer scientists, and meteorologists and implemented an unprecedented array of sensors to gather 468 million depth measurements, compared with 564 data points in previous models.

Analyzing this amount of data requires complex multiscalar data architectures, significant computing resources, or software services (e.g., data storage). Automated decision-making systems and integrated big-data architecture standards are being developed but do not yet oper-

5. Smart Earth content cloud. Courtesy of authors.

ate seamlessly or with full interoperability.[13] Moreover, multiplying real-time data streams and interoperable technologies does not necessarily lead to more efficient or transparent modes of environmental governance. Indeed, enhanced data collection raises significant ethical issues (e.g., privacy) and security concerns, and it runs the risk of data colonialism.[14]

Once these significant data issues are addressed, new pathways in environmental governance frameworks will emerge. For example, consider the case of Sustainability Standards Organizations (ssos). Emergent modes of access to real-time, continuous environmental data are challenging the conventional sso model.[15] In the past sso audits were conducted through brief, intermittent field visits by small teams of auditors and experts. Digital technologies such as embedded sensors and blockchain algorithms now enable the continuous monitoring and assessment of sustainability claims. This enables the emergence of private regulatory bodies and real-time auditing processes, effectively transforming the role of an SSO. This is an example of a novel coproduction of technology and governance, which is likely to be more prevalent in the future.

These digitally powered environmental governance models will rely increasingly on algorithms. Victor Galaz and Abdul Mouazen point to a key challenge in this regard: the most powerful algorithms underlying automated decision-making systems are likely to be of limited accessibility or transparency, raising the risk of reproducing biases in decision-making.[16] As many scholars have pointed out, approaches driven by AI and machine learning often contain implicit bias; incomplete datasets and flawed algorithms can prove counterproductive.[17] Algorithms cannot reconcile value-laden tensions between competing uses and ecosystem services (e.g., economic versus spiritual values), leaving questions of arbitration unresolved. Perhaps counterintuitively, this creates a need for new forms of human supervision as the automation of decision-making intensifies.[18]

A related issue complicating the uptake of Smart Earth data in environmental governance is the lack of or variability in data standards. This problem can both support and further complicate appeals for cross-disciplinary collaboration on data standards and data sharing.[19] "Big ecology" policies, combined with diminished costs for information technologies and new cloud-based data archiving tools and repositories, are proliferating, supported by an increasing number of ecoinformatics scientists and organizations such as the Ecological Society of America's Committee on the Future of Long-Term Ecological Data.[20] Scholars of ecological informatics predict that the new generation of data-sharing networks will grow exponentially faster than its predecessors and in a more collaborative manner.[21] For example, it is increasingly common for large-scale observatories—whether underwater (e.g., VENUS) or in the sky (ARGOS)—to host multiple sensor arrays for multiple research communities.[22]

In summary, much of the ecoinformatics literature focuses on addressing data gaps, the need for more collaborative data sharing, and issues relating to data quality. The underlying assumption is that more comprehensive and higher-quality data will lead to more effective environmental governance. Analyses conducted by science and technology studies scholars demonstrate, however, that this assumption is problematic.[23] Political commitments associated with measurement—in particular, the question of who selects what variables to measure and for whom they are selected—are rarely theorized or discussed in most of the Smart Earth research we surveyed (although this is changing across forums of science and technology studies and geography in particular). Nevertheless, because data gaps are likely to become more acute, the problem of "what is measured matters" or "what is counted counts," and by extension "what is not counted doesn't," is likely to intensify across a range of Smart Earth applications.

Open-Source Data

A Smart Earth world abounds in rapidly circulating and messy open-source data. In the ecological sciences there is a growing (albeit contested) conviction that such data will enhance biodiversity conservation, even if it requires "scrubbing" (e.g., quality control).[24] Some scholars argue that open-source data is a necessary, though not sufficient, condition for customizing environmental governance at local scales, ensuring a more democratic model of nested environmental decision-making. At their most utopian, open-source advocates promise that ubiquitously available data can unite environmental governance and democratic rule as a matter of good practice.

There have been numerous efforts to survey the array of open-source archives being promoted within governments, research projects, and NGOs.[25] William Michener examines how "big data" information policies have encouraged the present deluge of open-source data, noting in particular the importance of long-term research networks, ecological observatory networks, and coordinated distributed experiments and observations networks.[26] Scientists have long used nonprofessionals, such as citizen scientists and community groups, to gather ecological information. But the apparent successes of particular well-publicized projects have been scrutinized and in some cases refuted.[27] Transnational data-sharing collaborations are not necessarily "collaborative" or characterized by "sharing." Indeed, these features have long been a horizon of possibility, but one that is abstracted from actual manifestations of Smart Earth.[28] A commonly cited issue is the absence of robust institutional support for collaborative data sharing, which could theoretically incentivize a culture of sharing. Critiques of crowdsourcing as a mechanism that tends to reinforce expert hierarchies point to a related problem in institutional adoptions of open source.[29]

Governing with open source leads again to the question of data standards, an issue with far-reaching consequence for environmental governance. Dominique Roche and colleagues surveyed one hundred datasets associated with studies in journals that commonly publish ecological and evolutionary research, finding that upward of 56 percent of the articles were linked to incompletely archived datasets.[30] Calls for improved metadata have become common in ecology and biodiversity science, even as "good news narratives" about the purported benefits of open source continue to obscure questions of quality and accountability.[31] The question of what constitutes an acceptable threshold for data quality prefigures a growing debate over the legitimacy of open-source data.

It has now been seven years since Mark Andrejevic and Mark Burdon's prognosis of a "sensory society" defined by a "growing array of networked digital devices" that "passively collect enormous amounts of data."[32] While we might debate the degree to which this dystopic vision has been realized, is it also clear that many of the world's existing environmental data infrastructures would not meet the criteria set forth by Lindsey Dillon and colleagues for "environmental data justice," usefully defined in one instance as "community-based environmental data collection, public (especially online) accessibility of environmental data, and environmental data platforms supported by an open-source online infrastructure—in particular, one that can be used and modified by local communities."[33] The degree to which environmental data infrastructures will undergo the "splintering" that Steve Graham and Simon Marvin note of urban systems is an open question.[34] Geoffrey Bowker once observed that the efforts to produce "global panopticons" of internally consistent, long-term environmental datasets are rarely achieved and are usually far more messy, less reproducible exercises than commonly understood.[35]

Today there are indeed many published reports bemoaning incomplete, disarticulated, or structurally complicated datasets and describing efforts to overcome these problems.[36] At a time when increasing amounts of data freely circulate and many scientists advocate for the continued diversification of research models, the political integration of diverse data streams remains largely unexplored terrain.[37] In coming to recognize what Shoshana Zuboff calls the "right to sanctuary"— namely, that social and spatial justice now consists, partially, in being defined by the *incompleteness* or *absence* of one's reported dataset—we identify one path (albeit one not necessarily accessible or available to all) toward more progressive cultures of environmental data.[38]

Prediction

Prediction has garnered considerable attention in the remote-sensing community, where new measurement techniques and modeling and visualization capacities are being used for a broad range of purposes, from forecasting weather patterns, earthquake events, and forest fires to biodiversity estimates.[39] Given the massive economic demands for environmental risk mitigation, it is not surprising to see considerable research in this area. Yet despite the promise of such predictive governance projects, their actual success rates have been questioned. Predictive determinations have to contend with socioeconomic and political variability, such as price fluctuations, uneven consumption patterns, and market crashes, each of which has powerful and idiosyncratic ecological effects.[40] Predictive capacities are further limited by the present-day assumptions of their programmers, who are often unable to internalize challenges that arise outside their framing contexts. Sean Cubitt's remark that "databases predict the predictable" is a reminder that certain ecological forms and processes may be excluded under automated tracking systems.[41]

The variability of spatial and temporal scales at which digital technologies are used to measure and monitor ecosystem processes is another challenge to robust predictive capacity. Complex ecosystems are characterized by distinct, variable spatial and temporal scales. For each system, there are critical and perhaps unique temporal and spatial scales at which sampling and monitoring need to occur. At times these scales may be overlaid and interact. For example, the relationship between climate and ocean circulation exhibits patterns on interannual to decadal timescales, whereas models of fish population dynamics often need to additionally include short-term variables that influence fish survival rates. Thus in some cases we do not yet have a sufficiently developed understanding of ecosystems to determine which variables must be monitored and on which temporal and spatial scales.

Notwithstanding these critiques, ecological change increasingly appears as knowable and hence programmable in a Smart Earth world.[42] This emphasis on programmability as an inherent characteristic of predictable ecosystems builds on research in adaptive monitoring, adaptive management, and anticipatory governance. Many studies provide examples of enhancements to predictive capacities enabled by new web platforms (e.g., OakMapper, Global Fishing Watch, Global Forest Watch). Through such platforms, managers and scientists can source data inputs for environmental niche models and predict risks, disturbances, and terrestrial transformations, as well as illegal fishing, logging, or poaching.

In summary, digital technologies enable new modes of prediction that create new possibilities for environmental governance. Adaptation and anticipation are now topics of considerable institutional interest at the national and supranational scale, where they inform debates over what Martin Mahony calls "the predictive state."[43] This is part of a longer history beyond the scope of this chapter, but it should be noted that systems of predictive analysis—from weather forecasts to energy forecasts—have tended to preempt political discussion on the basis of scientific objectivity. Scholars point to proliferating ethical dilemmas—including questions of privacy, freedom, and security—posed by the emerging predictive governance systems, which are increasingly dependent (however imperfectly) on ubiquitous surveillance systems, some of which promise monitoring at a planetary scale.[44]

Real-Time Regulation

The concept of real time—for example, the increasingly instantaneous actual time elapsed in the performance of a computation—is central to Smart Earth governance.[45] In the last decade a rapid decline in the cost of monitoring technologies, driven by innovations in engineering, computing, and communications, has increased the capacity to conduct near real-time assessment of environmental changes.[46] Managers are evaluating the success of real-time location information via software applications on location-aware devices. Scholars and state planners have proposed the implementation of real-time responses to hazards such as earthquakes and fires, as well as real-time regulation of resource extractive sectors. In response to accelerating climate destabilization, both reforms in governance frameworks and the pace of governance will similarly accelerate.

A paradigmatic instance of real time is the work of Conserve.io, which assists conservation groups with leveraging mobile, cloud-based, and big-data technologies through distributed data collection. The mobile applications produced by Conserve.io use visual analytics to enhance situational awareness and fast responses to environmental changes.[47] For example, Whale Alert gathers real-time marine mammal sightings in marine shipping zones and uses the data to alert ship captains of unsafe vessel proximities, with the goal of reducing ship strikes.[48]

Whereas management and critical policy literatures focus on real-time resource distribution concerns—for example, coordinating flows of energy, bodies, and commodities—ecologists and biologists have been more engaged in novel species tracking efforts.[49] It is now possible to distill near real-time patterns from animal data flows. New technologies, such as conservation drones, augmented virtual environments, and conservation apps, are being proposed as adaptive solutions to monitoring and enforcement challenges.[50]

The challenge—and opportunity—posed by managing real-time data streams is one of the most salient issues for Smart Earth environmental governance. Real-time regulation poses significant administrative challenges, as organizing multiple temporal attributes or tracking efforts—including time of acquisition, integration/dwell time, sampling interval, and aggregation time span—can easily overwhelm computing capacities or lead to insufficient data. For an assailed species, the contingencies of a distant bandwidth connection may suddenly pose a material risk. It is likewise impossible to guarantee that ecological well-being will motivate the formation of a real-time response. New capacities in this area may lead to exploitation of natural resources as much as it does their conservation.

A second, related issue is the logistical challenge of sharing insights across spatially and institutionally distributed communities. Peter Fox and James Hendler observe a growing mismatch between the resource cost of creating scientific visualizations and the more rapidly decreasing costs of data generation (per unit of data generated).[51] Because digital devices are functional only to the extent that they are integrated into superannuating real-time networks, the problem of "dark data"—that is, data rendered invisible and hence unusable in Smart Earth practices—looms.[52] Real-time regulation presupposes a constant availability of power sources for its effective operation, an expectation that ignores the effects of climate change on data infrastructures.[53] Real-time distributional challenges could easily multiply in the face of a brownout or similar disruption. Cubitt's claim that insufficient electricity, not oil, is the leading threat to coordination of global governance has considerable salience for Smart Earth.[54] More research is needed on the implications for environmental governance of small but cascading energy stoppages, even at the orders of seconds and nanoseconds, which could inhibit the reliability of real-time regulation.[55]

Conclusions

Smart Earth technologies create conditions for significant shifts in environmental governance. Here we briefly summarize key points and areas for future research. First we should emphasize that better data does not necessarily lead to better governance. We have suggested that algorithms can selectively reduce the sphere of possible intervention and analysis within a particular setting. Mitigating against this tendency and demanding truly progressive principles *alongside* the maintenance of healthy ecosystems will likely require multiple and overlapping systems of oversight and quality control, with inevitable and ongoing political negotiation. Comprehensive analyses of regulatory gaps must be continuous and include close scrutiny of the integrative architectures proposed as global frameworks for storing, analyzing, and disseminating digital environmental data.

Another key research need is a critical analysis of the role of the state, historically a key player in multiscalar processes of environmental change. Future work needs to evaluate the role of the state in supporting or limiting Smart Earth processes at different moments. Smart Earth creates not only new ways of sensing and administering environments but also new categories of environmental assets. The chapters in part 3 of this volume explore these issues and consider the political and socioeconomic aspects of state engagement in adaptive management as digital technologies become more widespread in multiscalar environmental governance.

Questions of equitable access also merit more scrutiny. Smart Earth implies a shift to increasingly automated governance. This is an especially significant development for those without smartphones or other such devices. Certain actors will have a diminished voice in the Smart Earth world if they cannot "register as digital signals."[56] Such inequalities may become entrenched, exacerbated, and even weaponized through iterative practices of Smart Earth governance.

Last but not least, e-waste will be a significant issue.[57] Smart data is derived from an expanded array of material objects, continuously sampling and leaving material "externalities" in the physical world. Data processing will in turn require real-time big-data analytics with greater energy demands. E-waste will also pose new ecological problems for system managers and government institutions. Considerable problems are inherent in the Smart Earth proliferation of screen-based technologies, owing to their externalities and disposal challenges. Innovations in batteries, power-saving technologies, and backups will be essential to the functioning and performance of actually existing smart grids, app-based conservation efforts, and the like.

Given these concerns, Galaz and Mouazen are justified in calling for a code of conduct (a "biosphere code") that allows people to take stock of the new social relationships and ethical challenges created by Smart Earth forms of governance.[58] Data-sharing policies and ecological measurement standards require new forms of visibility in public debate. The demands of differently abled bodies must be valued alongside those a "sensory society" claims to "see." Sheila Jasanoff's demand for "technologies of humility" continues to resonate in a Smart Earth world.[59] Ethics is not an afterthought but a crucial foundation for the rapidly shifting governance of our digitizing planet.

Notes

1. Gale, Ascui, and Lovell, "Sensing Reality?"; Hampton et al., "Big Data"; Koomey, Matthews, and Williams, "Smart Everything"; Michener and Jones, "Ecoinformatics"; Pettorelli et al., "Satellite Remote Sensing."

2. Cohen and McCarthy, "Reviewing Rescaling"; Reed and Bruyneel, "Rescaling Environmental Governance."

3. Adapted from Bakker and Ritts, "Smart Earth."

4. Jepson and Ladle, "Nature Apps"; Joppa, "Case for Technology Investments."

5. Lindenmayer, Likens, and Franklin, "Earth Observation Networks."

6. Matabos et al., "Expert, Crowd, Students or Algorithm."

7. Luque, McFarlane, and Marvin, "Smart Urbanism."

8. Data One, https://www.dataone.org/; Anand, Gupta, and Appel, *Promise of Infrastructure*.

9. Thatcher, O'Sullivan, and Mahmoudi, "Data Colonialism through Accumulation"; Alvarez León, "Property Regimes."

10. Goodchild and Glennon, "Crowdsourcing Geographic Information"; Resch et al., "Pervasive Geo-security"; Koomey, Matthews, and Williams, "Smart Everything."

11. Arts, van der Wal, and Adams, "Digital Technology"; Cubitt, *Finite Media*; Gabrys, *Program Earth*; Swanstrom, *Animal, Vegetable, Digital*.

12. Gilbert, "Inner Workings."

13. Thatcher, "Big Data, Big Questions."

14. Thatcher, O'Sullivan, and Mahmoudi, "Data Colonialism through Accumulation."

15. Gale, Ascui, and Lovell, "Sensing Reality?"

16. Galaz and Mouazen, "'New Wilderness.'"

17. Caliskan, Bryson, and Narayanan, "Semantics Derived Automatically."

18. Galaz and Mouazen, "'New Wilderness,'" 629.

19. Frew and Dozier, "Environmental Informatics"; Hampton et al., "Big Data"; Michener, "Ecological Data Sharing."

20. Porter, "Brief History of Data Sharing."

21. Michener, "Ecological Data Sharing"; Reichman, Jones, and Schildhauer, "Challenges and Opportunities"; Zimmerman, "New Knowledge."

22. Starosielski, *Undersea Network*; Benson, "One Infrastructure."

23. Gabrys, "Programming Environments"; Gabrys, *Program Earth*; Jasanoff, "Technologies of Humility."

24. Hemmi and Graham, "Hacker Science"; Morris and White, "EcoData Retriever"; Turner et al., "Free and Open-Access Satellite Data."

25. Gale, Ascui, and Lovell, "Sensing Reality?"; Rocchini et al., "Open-Access and Open-Source"; Roche et al., "Public Data Archiving"; Welle Donker and van Loenen, "How to Assess?"

26. Michener, "Ecological Data Sharing."

27. See Blair et al., "Data Science."

28. Reichman, Jones, and Schildhauer, "Challenges and Opportunities"; Volk, Lucero, and Barnas, "Why Is Data Sharing?"

29. Gabrys, *Program Earth*; Pearson et al., "Can We Tweet?"; Swanstrom, *Animal, Vegetable, Digital.*

30. Roche et al., "Public Data Archiving."

31. Frew and Dozier, "Environmental Informatics"; Specht et al., "Data Management Challenges"; O'Brien, Costa, and Servilla, "Ensuring the Quality"; Arts, van der Wal, and Adams, "Digital Technology," 661.

32. Andrejevic and Burdon, "Defining the Sensor Society," 19.

33. Dillon et al., "Environmental Data Justice," 187.

34. Graham and Marvin, *Splintering Urbanism.*

35. Bowker, "Biodiversity, Datadiversity."

36. Hermes, Pearlman, and Buttigieg, "What's the Best Way?"

37. Alvarez León, "Property Regimes"; Verburg et al., "Methods and Approaches."

38. Zuboff, *Age of Surveillance Capitalism*, 477.

39. Castelli, Vanneschi, and Popovič, "Predicting Burned Areas"; Mairota et al., "Challenges and Opportunities"; Murai, "Can We Predict Earthquakes?"; Rocchini, Hernández-Stefanoni, and He, "Advancing Species Diversity Estimate"; Woodward, Gray, and Baird, "Biomonitoring."

40. Moore, *Capitalism in the Web of Life.*

41. Cubitt, *Finite Media*, 159.

42. Frew and Dozier, "Environmental Informatics"; Gabrys, *Program Earth*; Leszczynski, "Speculative Futures"; Murai, "Can We Predict Earthquakes?"

43. Mahony, "Predictive State."

44. Wood, "What Is Global Surveillance?"

45. De Longueville et al., "Digital Earth's Nervous System."

46. Koomey, Matthews, and Williams, "Smart Everything?"

47. Conserve.io, http://conserve.io/.

48. Whale Alert, http://www.whalealert.org/.

49. Benson, "One Infrastructure"; Gabrys, *Program Earth.*

50. Jepson and Ladle, "Nature Apps"; Jian et al., "Augmented Virtual Environment"; Sandbrook, "Social Implications."

51. Fox and Hendler, "Changing the Equation."

52. Wilson, "Continuous Connectivity"; Hampton et al., "Big Data"; Roche et al., "Public Data Archiving."

53. Durairajan, Barford, and Barford, "Lights Out."

54. Cubitt, *Finite Media.*

55. Durairajan, Barford, and Barford, "Lights Out."

56. Crawford, Gray, and Miltner, "Critiquing Big Data," 1667.

57. Cubitt, *Finite Media.*

58. Galaz and Mouazen, "'New Wilderness,'" 629.

59. Jasanoff, "Technologies of Humility."

Bibliography

Alvarez León, Luis. "Property Regimes and the Commodification of Geographic Information: An Examination of Google Street View." *Big Data & Society* 3, no. 2 (2016): 1–13.

Anand, Nikhil, Akhil Gupta, and Hannah Appel. *The Promise of Infrastructure.* Durham NC: Duke University Press, 2018.

Andrejevic, Mark, and Mark Burdon. "Defining the Sensor Society." *Television & New Media* 16, no. 1 (2015): 19–36.

Arts, Koen, René van der Wal, and William Adams. "Digital Technology and the Conservation of Nature." *Ambio* 44, no. 4 (2015): 661–73.

Bakker, Karen, and Max Ritts. "Smart Earth: A Meta-Review and Implications for Environmental Governance." *Global Environmental Change* 52 (2018): 201–11.

Benson, Etienne. "One Infrastructure, Many Global Visions: The Commercialization and Diversification of Argos, a Satellite-Based Environmental Surveillance System." *Social Studies of Science* 42, no. 6 (2012): 843–68.

Biermann, Frank, Kenneth Abbott, Steinar Andresen, Karin Bäckstrand, Steven Bernstein, Michele Betsill, Harriet Bulkeley et al. "Navigating the Anthropocene: Improving Earth System Governance." *Science* 335, no. 6074 (2012): 1306–7.

Blair, Gordon S., Peter Henrys, Amber Leeson, John Watkins, Emma Eastoe, Susan Jarvis, and Paul J. Young. "Data Science of the Natural Environment: A Research Roadmap." *Frontiers in Environmental Science* 7, no. 121 (2019).

Bowker, Geoffrey. "Biodiversity, Datadiversity." *Social Studies of Science* 30, no. 5 (2000): 643–83.

Bridge, Gavin, and Tom Perreault. "Environmental Governance." In *A Companion to Environmental Geography,* edited by Castree Noel, David Demeritt, Diana Liverman, and Bruce Rhoads, 475–97. West Sussex: Malden, 2009.

Caliskan, Aylin, Joanna Bryson, and Arvind Narayanan. "Semantics Derived Automatically from Language Corpora Contain Human-Like Biases." *Science* 356, no. 6334 (2017): 183–86.

Castelli, Mauro, Leonardo Vanneschi, and Aleš Popovič. "Predicting Burned Areas of Forest Fires: An Artificial Intelligence Approach." *Fire Ecology* 11, no. 1 (2015): 106–18.

Cohen, Alice, and James McCarthy. "Reviewing Rescaling: Strengthening the Case for Environmental Considerations." *Progress in Human Geography* 39, no. 1 (2015): 3–25.

Crawford, Kate, Mary Gray, and Kate Miltner. "Critiquing Big Data: Politics, Ethics, Epistemology; Special Section Introduction." *International Journal of Communication* 8 (2014): 1663–72.

Cubitt, Sean. *Finite Media: Environmental Implications of Digital Technologies.* Durham NC: Duke University Press, 2016.

De Longueville, Bertrand, Alessandro Annoni, Sven Schade, Nicole Ostlaender, and Ceri Whitmore. "Digital Earth's Nervous System for Crisis Events: Real-Time Sensor Web Enablement of Volunteered Geographic Information." *International Journal of Digital Earth* 3, no. 3 (2010): 242–59.

DeLoughrey, Elizabeth. "Satellite Planetarity and the Ends of the Earth." *Public Culture* 26, no. 2 (2014): 257–80.

Dillon, Lindsey, Dawn Walker, Nicholas Shapiro, Vivian Underhill, Megan Martenyi, Sara Wylie, Rebecca Lave, Michelle Murphy, Phil Brown, and EDGI. "Environmental Data Justice and the Trump Administration: Reflections from Environmental Data and Governance Initiative." *Environmental Justice* 10, no. 6 (October 2017): 186–92.

Durairajan, Ramakrishnan, Carol Barford, and Paul Barford. "Lights Out: Climate Change Risk to Internet Infrastructure." *Proceedings of the Applied Networking Research Workshop* (2018): 9–15.

Fox, Peter, and James Hendler. "Changing the Equation on Scientific Data Visualization." *Science* 331, no. 6018 (2011): 705–8.

Frew, James, and Jeff Dozier. "Environmental Informatics." *Annual Review of Environment and Resources* 37 (2012): 449–72.

Gabrys, Jennifer. *Program Earth: Environmental Sensing Technology and the Making of a Computational Planet.* Minneapolis: University of Minnesota Press, 2016.

———. "Programming Environments: Environmentality and Citizen Sensing in the Smart City." *Environment and Planning D: Society and Space* 32, no. 1 (2014): 30–48.

Galaz, Victor, and Abdul Mouazen. "'New Wilderness' Requires Algorithmic Transparency: A Response to Cantrell et al." *Trends in Ecology & Evolution* 32, no. 9 (2017): 628–29.

Gale, Fred, Francisco Ascui, and Heather Lovell. "Sensing Reality? New Monitoring Technologies for Global Sustainability Standards." *Global Environmental Politics* 17, no. 2 (2017): 65–83.

Gilbert, Natasha. "Inner Workings: Smart-Sensor Network Keeps Close Eye on Lake Ecosystem." *Proceedings of the National Academy of Sciences* 115, no. 5 (2018): 828–30.

Goodchild, Michael, and J. Alan Glennon. "Crowdsourcing Geographic Information for Disaster Response: A Research Frontier." *International Journal of Digital Earth* 3, no. 3 (2010): 231–41.

Graham, Steve, and Simon Marvin. *Splintering Urbanism: Networked Infrastructures, Technological Mobilities and the Urban Condition.* New York: Routledge, 2001.

Hampton, Stephanie, Carly Strasser, Joshua Tewksbury, Wendy Gram, Amber Budden, Archer Batcheller, Clifford Duke, and John Porter. "Big Data and the Future of Ecology." *Frontiers in Ecology and the Environment* 11, no. 3 (2013): 156–62.

Hemmi, Akiko, and Ian Graham. "Hacker Science versus Closed Science: Building Environmental Monitoring Infrastructure." *Information, Communication & Society* 17, no. 7 (2014): 830–42.

Hermes, Juliet, Jay Pearlman, and Pier Luigi Buttigieg. "What's the Best Way to Responsibly Collect Ocean Data?" *Eos* 99 (2018).

Jasanoff, Sheila. "Technologies of Humility: Citizen Participation in Governing Science." *Minerva* 41, no. 3 (2003): 223–44.

Jepson, Paul, and Richard J. Ladle. "Nature Apps: Waiting for the Revolution." *Ambio* 44, no. 8 (2015): 827–32.

Jian, Hongdeng, Jingjuan Liao, Xiangtao Fan, and Zhuxin Xue. "Augmented Virtual Environment: Fusion of Real-Time Video and 3D Models in the Digital Earth System." *International Journal of Digital Earth* 10, no. 12 (2017): 1177–96.

Joppa, Lucas. "The Case for Technology Investments in the Environment." *Nature* 552, no. 7685 (2017): 325–28.

Kitchin, Rob. "Big Data, New Epistemologies and Paradigm Shifts." *Big Data & Society* 1, no. 1 (2014): 2053951714528481.

Koomey, Jonathan, H. Scott Matthews, and Eric Williams. "Smart Everything: Will Intelligent Systems Reduce Resource Use?" *Annual Review of Environment and Resources* 38 (2013): 311–43.

Leszczynski, Agnieszka. "Speculative Futures: Cities, Data, and Governance beyond Smart Urbanism." *Environment and Planning A* 48, no. 9 (2016): 1691–708.

Lindenmayer, David, Gene Likens, and Jerry Franklin. "Earth Observation Networks (EONs): Finding the Right Balance." *Trends in Ecology & Evolution* 33, no. 1 (2018): 1–3.

Luque, Andrés, Colin McFarlane, and Simon Marvin. "Smart Urbanism: Cities, Grids and Alternatives?" In *After Sustainable Cities?*, edited by Mike Hodson and Simon Marvin, 74–89. London: Routledge, 2014.

Mahony, Martin. "The Predictive State: Science, Territory and the Future of the Indian Climate." *Social Studies of Science* 44, no. 1 (2014): 109–33.

Mairota, Paola, Barbara Cafarelli, Raphael Didham, Francesco Lovergine, Richard Lucas, Harini Nagendra, Duccio Rocchini, and Cristina Tarantino. "Challenges and Opportunities in Harnessing Satellite Remote-Sensing for Biodiversity Monitoring." *Ecological Informatics* 30 (2015): 207–14.

Matabos, Marjorie, Maia Hoeberechts, Carol Doya, Jacopo Aguzzi, Jessica Nephin, Thomas E. Reimchen, Steve Leaver, Roswitha Marx, Alexandra Branzan Albu, Ryan Fier, Ulla Fernandez-Arcaya, and S. Kim Juniper. "Expert, Crowd, Students or Algorithm: Who Holds the Key to Deep-Sea Imagery 'Big Data' Processing?" *Methods in Ecology and Evolution* 8 (2017): 996–1004.

Michener, William. "Ecological Data Sharing." *Ecological Informatics* 29 (2015): 33–44.

Michener, William, and Matthew Jones. "Ecoinformatics: Supporting Ecology as a Data-Intensive Science." *Trends in Ecology & Evolution* 27, no. 2 (2012): 85–93.

Moore, Jason. *Capitalism in the Web of Life: Ecology and the Accumulation of Capital.* Brooklyn: Verso, 2015.

Morris, Benjamin, and Ethan White. "The EcoData Retriever: Improving Access to Existing Ecological Data." *PLOS ONE* 8, no. 6 (2013): e65848.

Murai, Shunji. "Can We Predict Earthquakes with GPS Data?" *International Journal of Digital Earth* 3, no. 1 (2010): 83–90.

O'Brien, Margaret, Duane Costa, and Mark Servilla. "Ensuring the Quality of Data Packages in the LTER Network Data Management System." *Ecological Informatics* 36 (2016): 237–46.

Pearson, Elissa, Hayley Tindle, Monika Ferguson, Jillian Ryan, and Carla Litchfield. "Can We Tweet, Post, and Share Our Way to a More Sustainable Society? A Review of the Current Contributions and Future Potential of #Socialmediaforsustainability." *Annual Review of Environment and Resources* 41 (2016): 363–97.

Pettorelli, Nathalie, William Laurance, Timothy O'Brien, Martin Wegmann, Harini

Nagendra, and Woody Turner. "Satellite Remote Sensing for Applied Ecologists: Opportunities and Challenges." *Journal of Applied Ecology* 51, no. 4 (2014): 839–48.

Porter, John. "A Brief History of Data Sharing in the U.S. Long Term Ecological Research Network." *Bulletin of the Ecological Society of America* 91, no. 1 (2010): 14–20.

Reed, Maureen, and Shannon Bruyneel. "Rescaling Environmental Governance, Rethinking the State: A Three-Dimensional Review." *Progress in Human Geography* 34, no. 5 (2010): 646–53.

Reichman, James, Matthew Jones, and Mark Schildhauer. "Challenges and Opportunities of Open Data in Ecology." *Science* 331, no. 6018 (2011): 703–5.

Resch, Bernd, Bernhard Schulz, Manfred Mittlboeck, and Thomas Heistracher. "Pervasive Geo-security: A Lightweight Triple-A Approach to Securing Distributed Geoservice Infrastructures." *International Journal of Digital Earth* 7, no. 5 (2014): 373–90.

Rocchini, Duccio, José Luis Hernández-Stefanoni, and Kate He. "Advancing Species Diversity Estimate by Remotely Sensed Proxies: A Conceptual Review." *Ecological Informatics* 25 (2015): 22–28.

Rocchini, Duccio, Vaclav Petras, Anna Petrasova, Ned Horning, Ludmila Furtkevicova, Markus Neteler, Benjamin Leutner, and Martin Wegmann. "Open Data and Open Source for Remote Sensing Training in Ecology." *Ecological Informatics* 40 (2017): 57–61.

Roche, Dominique, Loeske E. B. Kruuk, Robert Lanfear, and Sandra Binning. "Public Data Archiving in Ecology and Evolution: How Well Are We Doing?" *PLOS Biology* 13, no. 11 (2015): e1002295.

Sandbrook, Chris. "The Social Implications of Using Drones for Biodiversity Conservation." *Ambio* 44, no. 4 (2015): 636–47.

Specht, Alison, Siddeswara Guru, Luke Houghton, Lucy Keniger, Patrick Driver, Euan Ritchie, Kaitao Lai, and A. Treloar. "Data Management Challenges in Analysis and Synthesis in the Ecosystem Sciences." *Science of the Total Environment* 534 (2015): 144–58.

Starosielski, Nicole. *The Undersea Network*. Durham NC: Duke University Press, 2015.

Swanstrom, Elizabeth. *Animal, Vegetable, Digital: Experiments in New Media Aesthetics and Environmental Poetics*. Tuscaloosa: University of Alabama Press, 2016.

Thatcher, Jim. "Big Data, Big Questions | Living on Fumes: Digital Footprints, Data Fumes, and the Limitations of Spatial Big Data." *International Journal of Communication* 8 (2014): 1765–83.

Thatcher, Jim, David O'Sullivan, and Dillon Mahmoudi. "Data Colonialism through Accumulation by Dispossession: New Metaphors for Daily Data." *Environment and Planning D: Society and Space* 34, no. 6 (2016): 990–1006.

Turner, Woody, Carlo Rondinini, Nathalie Pettorelli, Brice Mora, Allison Leidner, Zoltan Szantoi, Graeme Buchanan et al. "Free and Open-Access Satellite Data Are Key to Biodiversity Conservation." *Biological Conservation* 182 (2015): 173–76.

Verburg, Peter, John Dearing, James Dyke, Sander Van Der Leeuw, Sybil Seitzinger, Will Steffen, and James Syvitski. "Methods and Approaches to Modelling the Anthropocene." *Global Environmental Change* 39 (2016): 328–40.

Volk, Carol, Yasmin Lucero, and Katie Barnas. "Why Is Data Sharing in Collaborative Natural Resource Efforts So Hard and What Can We Do to Improve It?" *Environmental Management* 53, no. 5 (2014): 883–93.

Welle Donker, Frederika, and Bastiaan van Loenen. "How to Assess the Success of the Open Data Ecosystem?" *International Journal of Digital Earth* 10, no. 3 (2017): 284–306.

Wilson, Matthew. "Continuous Connectivity, Handheld Computers, and Mobile Spatial Knowledge." *Environment and Planning D: Society and Space* 32, no. 3 (2014): 535–55.

Wood, David Murakami. "What Is Global Surveillance? Towards a Relational Political Economy of the Global Surveillant Assemblage." *Geoforum* 49 (2013): 317–26.

Woodward, Guy, Clare Gray, and Donald Baird. "Biomonitoring for the 21st Century: New Perspectives in an Age of Globalisation and Emerging Environmental Threats." *Limnetica* 32, no. 2 (2013): 159–74.

Zimmerman, Ann. "New Knowledge from Old Data: The Role of Standards in the Sharing and Reuse of Ecological Data." *Science, Technology, & Human Values* 33, no. 5 (2008): 631–52.

Zuboff, Shoshana. *The Age of Surveillance Capitalism: The Fight for a Human Future at the New Frontier of Power.* New York: PublicAffairs, 2020.

FOUR

Data, Colonialism, and the Transformation of Nature in the Pacific Northwest

Anthony Levenda and Zbigniew Grabowski

This chapter examines settler colonial and capitalist transformations of nature enabled by novel data infrastructures vis-à-vis resurgent tribal political power in the U.S. Pacific Northwest's Columbia River Basin (CRB). To understand why the region is now pivoting to embrace data industries, one must understand three ongoing drivers of regional infrastructure development. First, ongoing settler colonial dispossession of land directly attacks traditional forms of data inscription and storage, which are encoded in memory, ritual, and relations with land and nonhumans and persist to this day as counterhegemonic practices. Second, contemporary systems of river management have produced and required new forms of data infrastructure, enabling technoscientific management of relational river systems. Finally, recent expansions of digital infrastructures such as fiber-optic cables, smart electric grids, and data centers, many of which serve as global data infrastructures, expand the need for existing hydroelectric dams and forms of river management rooted in settler colonial and capitalist modes of production.

Through an analysis of how these three forces operate, we illustrate how social, ecological, and technological changes are intertwined with data infrastructures. In the first case, the landscape and knowledge of Indigenous peoples have been coproduced since "time immemorial," a term referring to the ways that landscapes, bodies, ecologies, and culture defy historical periodization (coming from a time outside of memory).[1] The second force, based on settler colonial dispossession of land and resources, created a new political-economic system relying on the dispossession of knowledge, land, and identity, changing human-nature relations from systems of stewardship and coinhabitance to one that maximizes extraction. Here the state manages land, fisheries, and hydropower using bureaucratic infrastructures, a crucial part of which is constructing historical narratives and human habitation of the landscape and devising indicators of environmental quality and fish populations for infrastructure operators.[2]

To these two relatively well-studied dimensions of sociotechnical transformation, we contribute an analysis of the trajectories of industrialization producing data colonialism: how data infrastructures create, embed, and reproduce colonial structures. We position this turn toward data as nature and data as a material force as an attempt to reconcile the interests of large hydropower infrastructures with resurgent tribal political power. Simultaneously, new recreational economies, energy technologies like wind power, and ongoing crises in regional fisheries have placed novel demands on hydropower operators. In response, hydropower operators seek new markets, particularly for baseload power, and have turned to data centers as reliable power customers. Thus current forms of data colonialism stem from systems of power and data distribution, manufacturing, and rail transport, containing but transcending the high modern ideal of rational scientific management of land and ecosystems as resources to maximize capital production. We conclude by showing how data infrastructures remain a function of the politics involved in representing and managing the environment and—ultimately—in the production of nature.

Data Infrastructures and Social, Ecological, and Technological Change

Infrastructures enable flows of nature to, within, and from society. Indigenous knowledges coproduce landscapes, which themselves continue to serve as infrastructures. We begin by following Indigenous scholars and activists' arguments that Indigenous knowledge is inseparable from the environment.[3] For example, Leanne Simpson argues that learning is always connected to land and that land-based pedagogy can form the basis for Indigenous resurgence and culturally grounded decolonial political projects.[4] Studies of Indigenous knowledge systems also show the ongoing importance of Indigenous science and ways of knowing.[5] While this politics of knowledge is founded on an intimate connection with the environment, data is a missing category of analysis in many accounts. But data is not new to Indigenous knowledge production or information storage. What *counts* as data has changed under Western colonization, and conversely, purposeful ignorance of traditional knowledge continues to be used to undermine Indigenous cultural and political self-determination.[6] As Desi Rodriguez-Lonebear explains:

> Indigenous peoples were relentlessly empirical with advanced systems of knowledge. For Indigenous peoples, data were everywhere, and survival was often tied to one's ability to gather, analyse and share this knowledge. The winter counts by the Plains Indians are an example of the meticulous and methodological nature of Indigenous data. The Lakota, Blackfeet and other Plains tribes recorded winter counts on animal hides to enumerate important aspects of their world. These detailed counts included numbers of tribal citizens, allies, enemies, wild game, lodges and so on: histories and assemblages of data that were instruments of survival. . . . Another instance of Indigenous peoples' detailed data-keeping are the totem poles carved in the Pacific North-West. Totem poles document everything from family histories and tribal origin stories to achievements, marriages and land rights.[7]

While data is commonly framed in terms of the "digital revolution," it has been collected for centuries. Data has been stored and transmitted by infrastructures without digital or computational aspects, such as notches on a totem pole, handwritten census counts, marks on the landscape, and extensive oral testimonies. Recognizing Indigenous forms of data requires a rethinking of data in terms of the intimate labor of making and storing it, by whom, and for whom. This helps situate current movements for data sovereignty as a practice of decolonization among Indigenous peoples in settler colonial states.[8] Thinking of nature as the root of all data infrastructure grounds data in material reality and recognizes its importance in social relations of storing and sharing information.

Large-scale settler colonial state projects within the Columbia River Basin—including large dams (fig. 6), industrial complexes, nuclear power and weapons facilities, and regional power transmis-

6. Map of the Columbia River Basin showing the locations of dams along the Columbia, Snake, and many other rivers in the basin. The Dalles Dam (13), Grand Coulee Dam (4), and Bonneville Dam (14) are discussed in this chapter. The map shows the extensive dam development in the region, including ownership by the U.S. federal government; public utilities; state, provincial, or local governments; and private entities. Wikimedia Commons.

sion networks—have required the displacement of Indigenous peoples and knowledge systems. Studies of hydropower dams, for instance, show how planners sought to tame nature by urbanizing it, bringing water and power to cities.[9] Seeing dams as central to the development of the modern colonial state connects the socioecological impacts of altered river ecologies with cascading effects like the introduction of new invasive species, diseases, and water quality issues.[10] Just as with built infrastructures, data infrastructures have always been intertwined in infrastructural systems. Scholars increasingly recog-

Levenda and Grabowski

nize how large-scale data infrastructures themselves rework human-nature relations.

Attention to the broadband, cloud computing systems, data centers, digital compression techniques, and internet protocols that allow the circulation and storage of media worldwide—and the integration of these with existing infrastructure systems—is a new field of inquiry.[11] Understanding the recent proliferation of data infrastructures like server farms, fiber-optic cables, and smart grids in the Pacific Northwest requires an interdisciplinary lens that builds on digital geographies, political ecology, and science and technology studies (STS) to analyze interdependent socioecological and technological transformations. Through such an approach, we explore how data has been produced, recorded, circulated, accessed, and stored by the infrastructures that are part of the region's most prominent "environmental" issues.

The vignettes in the following sections illustrate three ways for understanding the role of data in regional transformation: Indigenous knowledges, extractive settler colonial bureaucracies, and new forms of digital data colonialism. While we treat each separately, these three forces continue to act in complex ways that we cannot comprehensively describe here. The CRB is marked by several socioenvironmental struggles related to use of the river for irrigation, fishing, navigation, power production, recreation, and cultural practices. It is no surprise that the region has been analyzed through numerous case studies to show how ongoing struggles over treaty interpretation and issues of sovereignty and self-determination are inextricably linked to water, land, salmon, and infrastructural futures. By highlighting these case studies on Indigenous knowledges, settler colonial extraction, and new forms of digital data colonialism, we contribute a critical approach to understanding how data infrastructure reshapes the material world.

The "Original Instructions" and Pre-Colonial Data Infrastructures

The Columbia River, known in the Sahaptin language as Nch'I Wana, or "the Big River," is a landscape that has been sacred to Indigenous peoples inhabiting the region since time immemorial.[12] This idea and experience of having an identity that goes "beyond" memory is difficult for settler colonial society to grasp, primarily because it cannot imagine a time outside of its own relevance and favors historical narratives centering its own arrival and conceptions of time (such as the temporal extension of terra nullius). As Vine Deloria Jr. explains, the logic of settler colonialism requires simplified understandings of human-nature relations, bright lines of temporally bounded belonging evidenced by oversimplified racist theories of human migration that justify its own dispossessive practices.[13] Interestingly, the racist and colonial narratives of arrival, such as the Bering land bridge hypothesis, have been invalidated by Western data itself as archaeological evidence continues to push back "recorded" dates of human occupation in the region.[14] However, scientific recognition continues to divorce the *experience* of being related to the land from its study, specifically through the reliance on a specific type of material data—archaeological evidence—used to legitimize or refute preexisting experiential accounts of belonging, a process that seeks to maintain Western data as the arbiter of legitimate knowledge.[15]

In contrast, Vine Deloria Jr., Billy Frank, and Steve Pavlik's *Indians of the Pacific Northwest* provides an experiential account of networked Indigenous societies throughout the region. These complex regional societies often crossed paths at permanent settlements and fishing sites along the Columbia, where the seasonal migrations of humans and fifty-plus-pound fish had been knit together since time immemorial.[16] Similarly, Winona LaDuke opens her book *All Our Relations* by describing the connections of peoples in the Pacific Northwest, including the Tillamook, Siletz, Yaquina, Alsea, Siuslaw, Umpqua, Hanis, Miluk, Colville, Tututni, Shasta, Costa, and Chetco, with salmon. She tells the story of the Tygh Band of the Lower Deschutes River in Oregon, who were struggling against dams that decimated salmon populations central to their way of life. She illustrates with a quote from a Tygh artist, fisherwoman, and community organizer who lamented the way non-Indigenous outsiders limit their environmental concerns to salmon counts: "The people are the salmon, and the salmon are the people. How do you quantify that?"[17] In this conception, salmon cannot be separated from their relations with Indigenous peoples or the cultural practices that make quantities and qualities of fish meaningful.

It is precisely because Indigenous data infrastructures do not require validation by colonial knowledge systems that they are so threatening to the settler colonial project. The concept of "original instructions" is a collection of stories and memories inscribed in the landscape and in the minds of Indigenous peoples, retaining information about the proper timing of ceremonies, notable historical events, and lessons that guide harmonious human and ecosystem relations. The original instructions are not merely a foil to colonial ways of knowing.[18] As researchers of European heritage, we do not claim intimate knowledge of these original instructions. Instead, by centering narratives produced by Indigenous scholars and activists, we wish to illustrate the significance of this way of being with the land for thinking about nature as data. It tells us something fundamental about the relationship between embodied knowledge and the landscape; this in itself is a form of data infrastructure.

Take, for instance, the work of Yakama scholars Virginia Beavert and Michelle Jacob, who document resurgent cultural practices in the region.[19] These rituals revitalize and reclaim the place-based specificity of Sahaptin languages, further supported by a Sahaptin dictionary and extensive language programs seeking to reengage youth with traditional culture.[20] These languages contain numerous terms that refer to specific plants and animals at specific times of the year in specific places.[21] Such an intimacy, encoded in place-based language as data, orders life outside of the bounds of time and space. This stands in stark contrast to Western systems of scientific classification, evidenced by the Linnaean species catalog and its contemporary derivative, the gene bank, which seeks to freeze biological diversity within an easily searchable form accessible to those without knowledge of these organisms in their cultural and ecological contexts.

In the CRB the diversity of cultural groups resists generalization. Rather, numerous interdependent Indigenous data systems occupy regionally specific zones. Yet these distinct knowledge systems overlapped and intersected through seasonal migrations and social gatherings, allowing for the exchange of goods, stories, and family relations.[22] More importantly, the region was thick with systems of customary law regulating trade, fishing activities, and social conflicts, especially in areas such as Celilo Falls, where diverse groups came together during salmon migrations.[23] Especially important was the knowledge of kin relations, allowing for access to hereditary fishing sites and the practices of distributing food to those who needed it the most. All this exhibits the markings of a place-based, relational, and intensive politics of care attacked by the expanding structures of settler colonialism seeking to reduce environmental management to a series of technologically constructed indicators.[24]

Legible Knowledge and the Settler Colonial State

It is tempting to reduce the settler colonial encounter to one of violence and dispossession. Yet such a narrative suppresses the power and military superiority of Indigenous peoples through most of the early period of the American settler colonial project. Early American settlements in the region were vulnerable, unlawfully ceded to the United States from the British, who had claimed the land (also illegally) after the War of 1812, which led U.S. president Pierce to give the Territory of Washington governor Stevens explicit instructions in 1853 to "make peace as quickly as possible" with basin tribes.[25] European settlement in the region also occurred outside of state-supported migration, as marginalized peoples fled state oppression and formed lawful cohabitation agreements, and even alliances, with Indigenous peoples against settler colonial states.[26]

While the idea of a settler colonial state is often interpreted through the lens of territorial claims, the diversity of the CRB shows us the contingency of Indigenous social relations that are legible to the settler colonial state as a form of data. In this sense the first settler colonial data infrastructure was the documentation of a set of tribal identities and political delineations, replete with censuses that oversimplified individual identities in order to assign them to specific reservations. These treaty negotiations, which took place between 1853 and 1855, are often referred to as the Stevens Treaties.[27] Similarly to Jacques Rancière's notion of politics as the designation of those who belong to the political class, and those who do not, the treaty process made certain actors representative of tribal identities. This was a process of double recognition, in that individuals were treated as representatives of their tribes, and bounded tribal identities made the treaty process more legible.

Levenda and Grabowski

These "politics of recognition" simplified dynamic patterns, such as the spatial complexity of the seasonal migration of Indigenous political units along rivers, into forms of data amenable to state negotiation and regulation of identity.[28]

Creating a written record of who served as "representatives" of specific tribes in order to delineate tribal lands thus represented the first move toward data colonialism, which continues to have ramifications to this day. Notably, many Indigenous peoples who refused to flatten their tribal identities into a single category also refused to move to the reservations.[29] Their interpretation of the oft cited "usual and accustomed places" included use of their traditional fishing sites along the rivers, which, as land became increasingly privatized and restricted under settler colonialism, also became sites of permanent habitation.[30] Denial of land access was a violation of treaty rights, as the Indigenous interpretation of treaties makes clear that treaty lands were to be held in a conditional trust (akin to a lien), meaning settlers' rights to occupy land were conditional on the United States upholding its treaty obligations. Settler systems of managing land, however, did not document customary uses that conflicted with settler colonial uses of land. These treaty violations, combined with settler militias' land claims made before treaty enactment, initiated several regional wars, which ceased only after the end of the Civil War allowed the United States to send its full military weight against regional tribes, as well as other Indigenous peoples throughout the West.[31]

Once human-human and human-land relations were made legible using settler colonial data infrastructures, another problem emerged for the settler colonial state. Put simply, Indigenous peoples could not be pacified by the confines of private property relations so long as customary uses of land were possible. As Leanne Simpson explains, "Colonial powers attacked virtually every aspect of our knowledge systems during the most violent periods of the past five centuries by rendering our spirituality and ceremonial life illegal, attempting to assimilate our children and destroy our languages through the residential school system, outlawing traditional governance, and destroying the lands and waters to which we are intrinsically tied. Our knowledge comes from the land, and the destruction of the environment is a colonial manifestation and a direct attack on Indigenous Knowledge and Indigenous nationhood."[32] During the late nineteenth century such attacks took the form of state-sanctioned violence to bar Indigenous peoples' access to the Columbia River and to deny claims of the damages wrought by rail infrastructure development, mining, ranching, agriculture, exploitative fisheries, and dams. Early settler colonial records show that this period of resource extraction, including near extermination of beaver for global fur markets, the draining of beaver ponds and other wetlands for agriculture, river modification for log transport, and extensive irrigation dam development, profoundly altered human-nature relations.

Crucially, there are limited records of the settler colonial state collecting data on key habitats and populations during this time, as the overall focus was on maximum exploitation. In some instances harvests and yields were tallied, such as estimates of annual salmon harvests. Yet a careful accounting that would support state-sponsored management of land, water, and life came only with the physical and bureaucratic infrastructures built during the New Deal and World War II.[33]

Dams, Data Centers, and the Future of a Region

Within the context of the CRB, the proliferation of modern data infrastructures was made possible by both the "nation-building" infrastructural projects of settler colonialism and the emergent economies of data use. Taken together, the legacies of river management for power, irrigation, and navigation—and the land use changes associated with them—are becoming more firmly intertwined with the use of computers and digital technologies locally and globally. Through their connections with existing hydropower infrastructures, rapidly proliferating data centers are expected to become a major user of electricity and water in consumer countries.[34]

In this sense the region has entered a new stage of contestation centered around its ability to provide infrastructural services to a global system of "data colonialism."[35] While much of the discussion on data colonialism has focused on the ways that user information is appropriated, the CRB shows how the physical colonial infrastructures are required by data-driven industries. It reminds us that the colonization

Levenda and Grabowski

of our data by extractive industries inextricably links our data production to the continued colonization of land and rivers. What the CRB makes clear is that mid-twentieth-century technoscientific management of landscapes through infrastructure, such as early computer systems used in hydropower operations to automate fish-counting systems, foreshadowed contemporary data technologies, such as smart grids and data centers.

The era of large dam building in the Pacific Northwest converted both the sacred landscape and the extractive infrastructures of settler colonialism to a utilitarian and engineered system, or that of an "organic machine."[36] The Bonneville Power Administration (BPA), U.S. Army Corps of Engineers (USACE), the Bureau of Reclamation (BOR), and a number of smaller regional utilities collaborated on basin-wide plans and led planning through the world war periods to extract power and facilitate flows of goods along the river.[37] This gave rise to a technomanagerial regime supported by data infrastructures that turned the river into a series of water stocks and flows, salmon counts, and megawatts of power for economic growth. Eleven major projects and hundreds of smaller dams were constructed during this time.[38]

The infrastructure for managing the basin (dams, reservoirs, and engineering knowledge) turned into a tool for land dispossession, water reallocation through irrigation and power programs, development of industrial logging and monocrop agriculture, and electricity-intensive industries like aluminum smelting. Such a technomanagerial regime required extensive organization of human labor as well as a rearrangement of land use within the basin to facilitate what Ashley Carse calls "nature as infrastructure."[39] As the river became infrastructure—a series of reservoirs, canals, and major dams like the Grand Coulee and Bonneville—it destroyed salmon spawning habitat and devastated native fisheries, burial and ceremonial sites, and farming lands. It was no accident that this era coincided with the criminalization of Indigenous fishing practices and overt attempts to follow through on policies of Indian removal along the river.

Throughout this settler colonial project of socioecological transformation, logics of "compensation" were put forth to address Indigenous claims to resources. Roberta Ulrich's essay "Empty Promises, Empty

Nets" provides a cutting account of dominant agencies' failure to live up to promises of compensation for dam building (an ongoing issue), which transformed the river into an engineered system.[40] The federal government promised to bring "in lieu" fishing sites to replace traditional fishing places behind the dams, such as the Dalles Dam in Oregon, which had "obliterated the most significant Native fishery on the Columbia River, and annihilated an economic and cultural place of several millennia."[41] Furthermore, Lindsey Schneider argues that limiting Native fishing through treaty rights can be thought of as the settler state's attempts to eliminate Native sovereignty and undermine place-specific decolonization strategies.[42]

One way that these conflicts were to be addressed was through the production of new data infrastructures to understand fish populations along the river, including fish ladder counts, fish transponder studies, and new ideas about how to keep fisheries from collapsing. Heather Anne Swanson maintains that these "often controversial scientific practices (such as fish counting) and analytics (such as carrying capacity) have served as tools not only for conservation advocacy but also, at times, for probing histories of settler colonialism and building alliances across difference."[43] With all this data available, Indigenous peoples have used their treaty rights to participate in salmon population science and data ownership as a way to address the oppression of settler colonialism. As Swanson notes, "While tribal leaders do not need fish population research to remind them of the ways that settler colonialism and industrial expansion have remade the Columbia Basin, they do seek its insights for how to tailor their practices of repair. . . . Salmon population numbers are a hard-won political accomplishment—a way of fighting to hold the state accountable for the harm it has wrought on the Columbia River Basin's fish."[44]

And yet these fish population counts are marked by their interdependence with dam infrastructure: they would not exist without the destructive dams. The official population counts of Columbia River salmon originating from the Bonneville Dam's fish ladder, which likely introduced significant bias into population estimates, occurred after severe overfishing and extensive habitat loss from smaller dams. Thus the "reliable" data of the scientific record used to set baseline population estimates guiding recovery targets is substantially less than traditional ecological knowledge of precolonial fish populations.

The transformation of the river into an energy generation infrastructure had other consequences for the river system, fish populations, and Indigenous peoples. Energy infrastructure requires demand, and the data behind projected demand was consistently used to boost regional economic development and industrialization. After World War II and into the Cold War era, hydroelectric power was consistently overproduced, and yet utilities used straight-line forecasts for demand growth based on historical data. This proved disastrous. Most visibly, the Washington Public Power Supply System (WPPSS, a consortium of electric utilities) projected such massive growth that it planned for five nuclear power plants in the region. With the 1970s energy crisis, however, price fluctuations and significant leveling of demand eventually led to a $2.25 billion municipal loan default because only one project was ever completed, while others were only partly constructed.[45]

This "WPPSS debacle" (pronounced locally as "whoops") led to the creation of the Northwest Power Act in 1980 to manage power planning and fish and wildlife conservation along the Columbia River System. Even after the Northwest Power Act guaranteed rate contracts for large industries, by 1996 the aluminum industry started to fall, decreasing power demand from the BPA contracts by 40 percent and continuing to decline ever since. The BPA has thus been searching for new sources of reliable electricity demand that match the needs of managing a massive hydropower system. At the same time, to reduce greenhouse gas emissions, federal policies pushing large-scale investments in wind power have introduced even greater uncertainty in regional power markets, causing the BPA to pay some wind power producers not to produce electricity in higher-water years.[46]

Today the BPA faces rising costs of producing power driven by underperformance, increasing maintenance costs, and legal obligations shaped by the Endangered Species Act, the Northwest Power Act, and the Columbia River Treaty between the United States and Canada. At the same time, the considerable amount of power they do produce has created new markets through data centers and the "wheeling" of electricity to California markets via the connections made between regional electricity grids.[47] With the proliferation of Google, Facebook, and Amazon data centers in the region—a good match for hydropower because of massive and reliable baseload electricity demands—new questions about power system management have arisen. In response to increasing power supply in the face of concomitant energy price declines and reduced demand driven by energy efficiency projects, and with ever more chaotic climatic patterns, the hydroelectric industry has started to decommission smaller hydroelectric dams, a trend likely to continue without greater electricity demand.[48] Thus the future of aging hydropower infrastructures has become uncertain in a system where their continued existence depends on a perception of their fiscal solvency, which increasingly relies on ensuring steady flows of electricity to massive data centers in the region.

The settler colonial transformation of the Columbia River into an engineered system providing cheap, often below-market rate power—combined with local tax incentives—has enabled data-intensive companies like Google, Facebook, and Amazon to locate and build in the Columbia River Basin.[49] At the same time, a narrative of regional economic development in which data industries drive economic growth, alongside logistics, software, and other "creative" and digital economy industries, generates new reasons for maintaining the Columbia power system, often in ways that marginalize Indigenous sovereignty. Of course, Native peoples are aware of this continued displacement of value and the uneven development that hydroelectric infrastructure has enabled. This is evidenced by statements made by a Yakama housing advocate at a public hearing in Dallesport about compensation for dam building and its associated infrastructure:

Native housing advocate: "We have persevered and kept a little bit of our ways even with desecration, profiling, trying to teach our kids our way, it's really hard for us every time we drive by the Dalles dam. I hope you appreciate every time you turn your lights on your children, your computer, your air conditioning, because you took that from our children."

Response from older white woman in crowd: "No we didn't!"

NHA: "Yes, you did."[50]

Thus the future of the region and the possibility of "just compensation" continue to be hotly debated, increasingly on the same digital platforms that are colonizing the region with their infrastructures. Current debates over managing the basin for fish, for capital, for cultural uses, and for emergent digital economies take place in this context. While the future is far from certain in a region built by federal investment, and thick with settler and Indigenous politics, we must remain aware of the power of data industries to coproduce regions through socio-eco-technical infrastructures.

Conclusions

This chapter has looked at how data infrastructures have coproduced socio-eco-technological transformations of the Columbia River Basin associated with settler colonialism and industrial capitalism. We traced complex entanglements of nature and data, from enduring stories encoded in the land to the technologies of technomanagerialism to the financial and sociotechnical impacts of global information technologies. These data-nature relationships have not displaced each other, but rather have become increasingly intertwined.

We have weighed how the river has been reshaped by human habitation, aspirations, systems of thought, science, government, and technology to become what Richard White dubbed "the organic machine." Such a framing of the river as a socio-eco-technological system remains just one way of seeing, knowing, and analyzing complex political, financial, environmental, social, and technological arrangements. It is not surprising that a river, much like the water that flows through it, continues to resist simplified systems of classification, categorization, and analysis, as it too is a living entity, quick to change but slow to evolve.

Presently, Indigenous activists take to Facebook to build social movements, Amazon and Google clandestinely fight for environmental deregulation, and local jurisdictions draw on treaty law to resist infrastructure expansion.[51] Within this increasingly complex set of relations, we are left with more questions about how nature and data are related under the expansion of data colonialism. As Lindsey Schneider writes in the context of battles over salmon and sea lions, "The colonization of the Columbia Basin drastically altered the ecological reality of the river, and at the same time, the river itself made that development possible. It simply does not make any sense to talk about the separation of nature and industry or to try to replicate the precolonial condition of just one species, aspect, or section of the river, when the river itself is a highly contested amalgam of ecological processes, Indigenous lifeways, and settler development."[52]

We need new questions about the value of data itself: What is the value of data and data infrastructures to different groups in the CRB? In the context of intensified data extractivism, regional infrastructures enabling capital accumulation, and ongoing battles over the future of the river and the region, how can we contribute to active decolonization? Indigenous peoples continue to fight for self-determination through the ongoing work of the Columbia Regional Intertribal Fish Commission and numerous resurgent cultural and political projects. What role will data and data centers play in the larger contestations over treaty rights and obligations? Our hope is that this chapter forges ideas for future scholarship connecting data infrastructure to broader socioecological transformations under data colonialism.

Notes

1. Deloria, Frank, and Pavlik, *Indians of the Pacific Northwest*; Deloria, *Red Earth, White Lies*.
2. Scott, *Seeing like a State*.
3. LaDuke, *All Our Relations*; Simpson, "Aboriginal Peoples and Knowledge."
4. Simpson, "Land as Pedagogy."
5. Deloria et al., "Unfolding Futures"; Smithers, "Beyond the 'Ecological Indian.'"
6. Deloria et al., "Unfolding Futures."
7. Rodriguez-Lonebear, "Building a Data Revolution," 254–55.
8. TallBear, "Genomic Articulations of Indigeneity."
9. Kaika, *City of Flows*.
10. Mitchell, *Rule of Experts*.
11. Parks and Starosielski, *Signal Traffic*; Pickren, "'Global Assemblage.'"
12. CRITFC, *Spirit of the Salmon Plan*.
13. Deloria, *Red Earth, White Lies*.
14. Davis et al., "Late Upper Paleolithic Occupation."
15. Stephens et al., "Archaeological Assessment."
16. Deloria, Frank, and Pavlik, *Indians of the Pacific Northwest*.
17. LaDuke, *All Our Relations*, 1–2.
18. CRITFC, *Spirit of the Salmon Plan*.
19. Beavert, *Gift of Knowledge*; Jacob, *Yakama Rising*.

Levenda and Grabowski

20. Beavert, *Gift of Knowledge*.

21. Hunn and Selam, *Nch'i-Wána, "the Big River."*

22. Hunn and Selam, *Nch'i-Wána, "the Big River"*; Fisher, *Shadow Tribe*.

23. Fisher, *Shadow Tribe*.

24. Fisher, *Shadow Tribe*.

25. Deloria, Frank, and Pavlik, *Indians of the Pacific Northwest*; Fisher, *Shadow Tribe*.

26. Grossman, *Unlikely Alliances*; Jacoby, *Crimes against Nature*.

27. Coulthard, "Place against Empire"; Fisher, *Shadow Tribe*.

28. Rancière, *Dissensus*; Fisher, *Shadow Tribe*.

29. Following Indigenous scholars such as Glen Coulthard, we use the term *category* here, as *construction* implies that these identities existed solely because of the activities of the colonizer rather than their own inherent multilayered logics.

30. CRITFC, *Spirit of the Salmon Plan*.

31. Fisher, *Shadow Tribe*.

32. Simpson, "Anticolonial Strategies," 377.

33. Netboy, *Salmon*.

34. Ristic, Madani, and Makuch, "Water Footprint."

35. Thatcher, O'Sullivan, and Mahmoudi, "Data Colonialism through Accumulation"; Couldry and Mejias, "Data Colonialism."

36. White, *Organic Machine*.

37. The BOR and USACE had historically competed in their basin-planning efforts until they agreed to coordinate their approaches for the Pick-Sloan plan for the Missouri River Basin, another famous example of dams displacing Indigenous peoples in the United States. See Worster, *Rivers of Empire*.

38. USACE's comprehensive river basin development planning was documented in the 308 Reports, otherwise known as the *Columbia River and Minor Tributaries*, updated in 1950 in House Document 531. These reports created a template and strategy for river basin management. For more information, see Molle, "River-Basin Planning."

39. Carse, "Nature as Infrastructure."

40. Ulrich, "Empty Promises, Empty Nets."

41. Robbins and Barber, *Nature's Northwest*, 154.

42. Schneider, "'There's Something in the Water.'"

43. Swanson, "Unexpected Politics of Population," 272.

44. Swanson, "Unexpected Politics of Population," 282.

45. Pope, *Nuclear Implosions*.

46. Pearsall, "Bonneville Power."

47. Bernton, "Bonneville."

48. Grabowski, Chang, and Granek, "Fracturing Dams, Fractured Data"; Grabowski et al., "Removing Dams, Constructing Science."

49. Levenda and Mahmoudi, "Silicon Forest and Server Farms."

50. Grabowski, "Transcript of Public Meeting."

51. Danko, "Board Upholds Denial."

52. Schneider, "'There's Something in the Water,'" 155.

Bibliography

Beavert, Virginia R. *The Gift of Knowledge / Ttnúwit Átawish Nch'inch'imamí: Reflections on Sahaptin Ways*. Seattle: University of Washington Press, 2017.

Bernton, Hal. "Bonneville, the Northwest's Biggest Clean-Power Supplier, Faces Promise and Perils in Changing Energy Markets." *Seattle Times*, July 21, 2019. https://www.seattletimes.com/seattle-news/bonneville-power-the-northwests-biggest-clean-energy-supplier-strains-to-keep-up-its-aging-hydro-system-and-restore-salmon/.

Carse, Ashley. "Nature as Infrastructure: Making and Managing the Panama Canal Watershed." *Social Studies of Science* 42, no. 4 (August 2012): 539–63.

Couldry, Nick, and Ulises A. Mejias. "Data Colonialism: Rethinking Big Data's Relation to the Contemporary Subject." *Television & New Media* 20, no. 4 (May 2019): 336–49.

Coulthard, Glen. "Place against Empire: Understanding Indigenous Anti-Colonialism." *Affinities: A Journal of Radical Theory, Culture, and Action* 4, no. 2 (November 2010): 79–83.

CRITFC (Columbia River Inter-Tribal Fish Commission). *Spirit of the Salmon Plan*. Portland OR: CRITFC, 2014.

Danko, Pete. "Board Upholds Denial of Union Pacific's Proposed Mosier Track Expansion." *Portland Business Journal*, June 14, 2017. https://www.bizjournals.com/portland/news/2017/06/14/board-upholds-denial-of-union-pacifics-proposed.html.

Davis, Loren G., David B. Madsen, Lorena Becerra-Valdivia, Thomas Higham, David A. Sisson, Sarah M. Skinner, Daniel Stueber et al. "Late Upper Paleolithic Occupation at Cooper's Ferry, Idaho, USA, ~16,000 Years Ago." *Science* 365, no. 6456 (August 2019): 891–97.

Deloria, Philip J., K. Tsianina Lomawaima, Bryan McKinley, Jones Brayboy, Mark N. Trahant, Loren Ghiglione, Douglas Medin, and Ned Blackhawk. "Unfolding Futures: Indigenous Ways of Knowing for the Twenty-First Century." *Daedalus* 147, no. 2 (March 2018): 6–16.

Deloria, Vine, Jr. *Red Earth, White Lies: Native Americans and the Myth of Scientific Fact*. Golden CO: Fulcrum Publishing, 1997.

Deloria, Vine, Jr., Billy Frank, and Steve Pavlik. *Indians of the Pacific Northwest: From the Coming of the White Man to the Present Day*. Golden CO: Fulcrum Publishing, 2016.

Fisher, Andrew H. *Shadow Tribe: The Making of Columbia River Indian Identity*. Seattle: University of Washington Press, 2010.

Grabowski, Zbigniew J. "Transcript of Public Meeting in Dallesport to Discuss Yakama Nation Housing Authority and US Army Corps of Engineers Proposal to Create Affordable Housing in Dallesport." June 19, 2018. Copy in the possession of the author.

Grabowski, Zbigniew J., Ashlie Denton, Mary Ann Rozance, Marissa Matsler, and Sarah Kidd. "Removing Dams, Constructing Science: Coproduction of Undammed Riverscapes by Politics, Finance, Environment, Society and Technology." *Water Alternatives* 10, no. 3 (2017): 769–95.

Grabowski, Zbigniew J., Heejun Chang, and Elise F. Granek. "Fracturing Dams, Fractured Data: Empirical Trends and Characteristics of Existing and Removed Dams in the United States." *River Research and Applications* 34, no. 6 (2018): 526–37.

Grossman, Zoltán. *Unlikely Alliances: Native Nations and White Communities Join to Defend Rural Lands*. Seattle: University of Washington Press, 2017.

Hunn, Eugene S., and James Selam. *Nch'i-Wána, "the Big River": Mid-Columbia Indians and Their Land*. Seattle: University of Washington Press, 1991.

Jacob, Michelle. *Yakama Rising*. Tucson: University of Arizona Press, 2013.

Jacoby, Karl. *Crimes against Nature: Squatters, Poachers, Thieves, and the Hidden History of American Conservation*. Berkeley: University of California Press, 2014.

Kaika, Maria. *City of Flows: Modernity, Nature, and the City*. New York: Routledge, 2012.

LaDuke, Winona. *All Our Relations: Native Struggles for Land and Life*. Chicago: Haymarket Books, 2017.

Levenda, Anthony M., and Dillon Mahmoudi. "Silicon Forest and Server Farms: The (Urban) Nature of Digital Capitalism in the Pacific Northwest." *Culture Machine* 18 (2019). https://culturemachine.net/vol-18-the-nature-of-data-centers/silicon-forest-and-server-farms/.

Mitchell, Timothy. *Rule of Experts: Egypt, Techno-Politics, Modernity*. Berkeley: University of California Press, 2002.

Molle, François. "River-Basin Planning and Management: The Social Life of a Concept." In "Gramscian Political Ecologies," edited by Michael Ekers, Alex Loftus, and Geoff Mann. Special issue, *Geoforum* 40, no. 3 (May 2009): 484–94.

Netboy, Anthony. *Salmon, the World's Most Harassed Fish*. Winchester ON: Winchester Press, 1980.

Parks, Lisa, and Nicole Starosielski. *Signal Traffic: Critical Studies of Media Infrastructures*. Urbana: University of Illinois Press, 2015.

Pearsall, Drew. "Bonneville Power Administration's Energy Curtailment Problem: An Analysis of Its Redispatch Policy and Oversupply Protocol P and Their Impact on Washington's Wind Power Producers, Utility Companies, and Energy Independence Act Notes & Comments." *Washington Journal of Environmental Law and Policy*, no. 1 (2013): 79–123.

Pickren, Graham. "'The Global Assemblage of Digital Flow': Critical Data Studies and the Infrastructures of Computing." *Progress in Human Geography* 42, no. 2 (April 2018): 225–43.

Pope, Daniel. *Nuclear Implosions: The Rise and Fall of the Washington Public Power Supply System*. New York: Cambridge University Press, 2008.

Rancière, Jacques. *Dissensus: On Politics and Aesthetics*. London: Bloomsbury Publishing, 2015.

Ristic, Bora, Kaveh Madani, and Zen Makuch. "The Water Footprint of Data Centers." *Sustainability* 7, no. 8 (August 2015): 11260–84.

Robbins, William G., and Katrine Barber. *Nature's Northwest: The North Pacific Slope in the Twentieth Century*. Tucson: University of Arizona Press, 2011.

Rodriguez-Lonebear, Desi. "Building a Data Revolution in Indian Country." In *Indigenous Data Sovereignty: Toward an Agenda*, vol. 38, edited by Tahu Kukutai and John Taylor, 253–72. Canberra, Australia: ANU Press, 2016.

Schneider, Lindsey. "'There's Something in the Water': Salmon Runs and Settler Colonialism on the Columbia River." *American Indian Culture and Research Journal* 37, no. 2 (January 2013): 149–64.

Scott, James C. *Seeing like a State: How Certain Schemes to Improve the Human Condition Have Failed*. New Haven CT: Yale University Press, 1998.

Simpson, Leanne. "Aboriginal Peoples and Knowledge: Decolonizing Our Process." *Canadian Journal of Native Studies* 21, no. 1 (2001): 137–48.

Simpson, Leanne Betasamosake. "Land as Pedagogy: Nishnaabeg Intelligence and Rebellious Transformation." *Decolonization: Indigeneity, Education & Society* 3, no. 3 (2014): 1–25.

Simpson, Leanne R. "Anticolonial Strategies for the Recovery and Maintenance of Indigenous Knowledge." *American Indian Quarterly* 28, no. 3/4 (2004): 373–84.

Smithers, Gregory D. "Beyond the 'Ecological Indian': Environmental Politics and Traditional Ecological Knowledge in Modern North America." *Environmental History* 20, no. 1 (January 2015): 83–111.

Stephens, Lucas, Dorian Fuller, Nicole Boivin, Torben Rick, Nicolas Gauthier, Andrea Kay, Ben Marwick et al. "Archaeological Assessment Reveals Earth's Early Transformation through Land Use." *Science* 365, no. 6456 (August 2019): 897–902.

Swanson, Heather Anne. "An Unexpected Politics of Population: Salmon Counting, Science, and Advocacy in the Columbia River Basin." *Current Anthropology* 60, no. S20 (June 2019): 272–85.

TallBear, Kim. "Genomic Articulations of Indigeneity." *Social Studies of Science* 43, no. 4 (August 2013): 509–33.

Thatcher, Jim, David O'Sullivan, and Dillon Mahmoudi. "Data Colonialism through Accumulation by Dispossession: New Metaphors for Daily Data." *Environment and Planning D: Society and Space* 34, no. 6 (December 2016): 990–1006.

Ulrich, Roberta. "Empty Promises, Empty Nets." *Oregon Historical Quarterly* 100, no. 2 (1999): 134–57.

White, Richard. *The Organic Machine: The Remaking of the Columbia River*. New York: Hill and Wang, 1995.

Worster, Donald. *Rivers of Empire: Water, Aridity, and the Growth of the American West*. Oxford: Oxford University Press, 1992.

Two

Civic Science and Community-Driven Data

FIVE

Environmental Sensing Infrastructures and Just Good Enough Data

Jennifer Gabrys and Helen Pritchard

Environmental data infrastructures typically operate at the levels of governments, institutions, and industry.[1] However, environmental data infrastructures are also increasingly materializing through citizen and community projects as people monitor environmental pollution, analyze data, and circulate findings in order to fight for environmental justice. The nature of data that the essays in this book investigate is transformed through these citizen practices, where the focus is on the effects and effectiveness of data and data practices as they are able to remedy or address environmental problems. Infrastructure in this sense consists not just of the devices that sense, gather, analyze, and communicate data but also of the community practices, relations, and modes of organizing that generate different ways of understanding and mobilizing environmental data for greater environmental and social justice.[2] The qualities and operations of environmental data change through these emerging infrastructures, including the way the data registers as significant within government, industry, and beyond.

In the context of these different environmental data infrastructures, this chapter considers the emergence of a number of environmental sensing technologies and practices. These initiatives seek to enable citizens to use do-it-yourself and low-tech monitoring tools to understand and act on environmental problems. One of the primary ways in which such "citizen sensing" projects have sprung up is through engagement with environmental pollution.[3] Noise, air, soil, and water pollution are local, if distributed, environmental disturbances that can now be monitored using a range of digital sensing devices.[4] These devices can provide different data in comparison to fixed and distant monitoring stations, while allowing citizen sensors to understand personal exposure more fully. A key motivating aim of many citizen sensing projects is to make the data gathered about environmental pollution a catalyst for political action. Such an objective represents a continuation of earlier citizen science initiatives that focused on gathering data or otherwise intervening within scientific practice to provide alternative forms of evidence based on diverse citizen experiences.[5]

In this way citizen-generated datasets are often gathered with equipment that diverges from state and regulatory standards and through practices that differ from standard scientific protocols. When monitoring air pollution or other environmental disturbances with low-cost technology, citizen-led initiatives are typically challenged about the validity or accuracy of their data. Environmental regulators at times dismiss citizen-collected data, since it is perceived to be biased, lacking in standardized procedures for collection, and generated through imprecise instruments. Yet citizens now deploy environmental-monitoring technologies in multiple contexts, and the amount and type of environmental data that they collect continues to grow. While citizen sensing technologies and practices might not typically involve consistently observing air pollution with sophisticated instruments to meet regulatory standards and ensure compliance with air pollution policy, they do involve capturing fine-grained pollution data through do-it-yourself devices that are often located in spatially dense networks and provide ongoing indications of relative changes in air quality, rather than absolute measurements.

As the U.S. Environmental Protection Agency (EPA) has expressed in its analysis of "next generation" environmental monitoring, "types of data" and "types of uses" are interlinked.[6] Data typically becomes admissible for legal claims only when gathered through specified scientific procedures by state-certified users with approved (and expensive) instrumentation. However, data gathered through citizen sensing practices may be "just good enough" in some situations for establishing that a pollution event is happening. It therefore remains a relatively open question as to what the uses and effects of data gathered through citizen sensing technologies might be, since these data practices are still emerging.

"Just good enough data" is a phrase and concept that we have devel-

Gabrys and Pritchard

oped to address issues of accuracy in citizen-collected data and expand the types of uses that might accompany these new types of data.[7] "Just good enough data" draws attention to attempts to counter the reliance on high levels of measurement accuracy as the sole criterion by which data is evaluated. What different practices emerge when engaging with environmental data in a more indicative, rather than regulatory, register? What do these practices enable? And what other relations, connections, and points of focus might "just good enough data" generate?

Examining these questions in the context of collaborative and participatory environmental sensing conducted by the Citizen Sense research group in northeastern Pennsylvania, this chapter considers how the use of air pollution monitors by residents living next to hydraulic fracturing, or fracking, infrastructure produces different registers and types of data.[8] We outline citizen sensing practices that monitor fracking-related pollution and discuss our attempts to contribute to these through further participatory and practice-based citizen sensing initiatives. The chapter reviews the multiple forms of data generated that diverge from state and regulatory monitoring, including air quality data, data logs, citizen observations and stories, and a tool developed by Citizen Sense to facilitate citizen-led analysis of data collected over nine months. It further discusses how residents attempted to mobilize data and engage in discussions with regulators and the ways that citizen-gathered data could provide other insights beyond a regulatory-only focus on monitoring.

Citizen-sensed data is rich with translocal experiences and collective insights, often bringing attention to environmental change from new perspectives. Rather than limit discussions of citizen sensing practices to accuracy and regulatory alignment, we investigate how to develop practices and infrastructures for "just good enough data" that enable citizen-sensed data to make expanded contributions to environmental sensing. We suggest that the relevance of citizen-collected air quality data is not solely determined through absolute criteria or alignment to state or federally managed air quality data, since these criteria can often shift depending on modes of governance, location, and available resources. If data is understood instead as a set of entities that transform depending on the uses to which they will be put—and how they might be "good enough" to achieve these ends—it then becomes possible to attend to how data is differently mobilized through monitoring practices and political encounters.[9] Emerging infrastructures can extend to data analysis and political organizing, as well as social connections and storying practices. These infrastructures contribute to distinct formations of "just good enough data."

Air Quality Monitoring and Natural Gas Extraction

Unconventional natural gas extraction in the form of hydraulic fracturing began in the Marcellus Shale region of Pennsylvania in 2003. However, by around 2006 the number of wells drilled in the state began to increase rapidly, and communities started to notice the impact of the industry more intensively.[10] As of November 2021 a total of 13,220 unconventional wells had been drilled as sites of natural gas production, with 17,968 violations logged at these well sites.[11] The rate of well development has continued to increase over the years. As recorded by a local citizen-led website that collates and provides information and data on gas well production in Pennsylvania, on average two new wells were permitted every day in just the three weeks preceding February 5, 2022.[12]

Many of these wells and the related natural gas infrastructure of compressor stations, well pads, glycol dehydrators, water impoundment ponds, and pipelines, are densely located in northeastern Pennsylvania. Along with this infrastructure, inevitable concerns have arisen about environmental impacts, especially in relation to water and air pollution. While much attention has been given to water pollution through several high-profile cases of contaminated well water, area residents have also had concerns about the relatively undermonitored effects of fracking on air quality.

Pennsylvania residents' interest in and sense of urgency about developing monitoring practices has also been in part a response to the lack of governmental monitoring in this rural area. Existing monitoring for the nationwide Air Quality Index (AQI), which is facilitated on a state level by the Pennsylvania Department of Environmental Protection (DEP), typically focuses more intensely on urban areas and roadside sites and does not pay particular attention to accounting for emissions from industries such as oil and gas. DEP stations for monitoring air quality are located in relatively distant urban centers such as Scranton, where monitors are often placed close to busy highways. As a result, pollutants such as particulate matter of 2.5 micrometers and smaller ($PM_{2.5}$) can go undetected in rural regions. Yet such pollutants are especially hazardous because of their small size and ability to circulate in the body and damage pulmonary and cardiac health.

Although the DEP also had undertaken some mobile monitoring on a sporadic basis, because of economic and political constraints no consistent monitoring by regulators had been taking place to more systematically account for local emissions from the natural gas industry in the northeastern part of the state. Within the context of a newly expanding industry that residents felt was not sufficiently monitored, interest in developing techniques for documenting environmental pollution in the area arose.[13]

To understand the air pollution arising from natural gas extraction and production, residents of Pennsylvania have undertaken many diverse practices of monitoring, with differing aims and objectives. Attempting to capture their experiences of air pollution and associated health effects, residents have used an extensive range of monitoring technologies either on their own or collaborating with or assisting scientific studies. Many monitoring practices have required that residents collect samples for lab analysis, which takes place in distant sites of data processing, or use technologies that produce data in forms that are not immediately usable or comparable with other datasets. The promise of low-cost and "next-generation" environmental sensors is that data will be made available in real time, in a legible output, to the users of the technologies.

A Participatory Approach to Citizen Sensing

By undertaking a collaborative and participatory approach to developing citizen sensing practices and technologies further with residents of northeastern Pennsylvania, the Citizen Sense research project held a series of discussions and monitoring events between 2013 and 2015.[14] Through this process, we developed the Citizen Sense Kit for air quality monitoring in the region (fig. 7). We deployed the kits to about thirty participants in October 2014, with a training workshop

7. Part of the Citizen Sense Kit. Courtesy of authors.

and walk to field-test the technologies. The research team then visited participants' homes to help set up the technologies, and participants developed a number of situations and experiments to monitor areas of particular concern to them.

The Citizen Sense Kit distributed during these events was developed in response to the concerns of community members, who provided information in logbooks that asked for input on the key concerns for natural gas infrastructure in relation to air pollution. The kit contained a passive sampling badge for monitoring emissions of benzene, toluene, ethylbenzene and xylene (BTEX), which are substances associated with gas production and that are also hazardous to human health, along with a Speck device from the CREATE Lab at Carnegie Mellon for monitoring $PM_{2.5}$.[15] The kit also included a custom-made prototype device, the Frackbox, which was installed at three compressor station sites. The Frackbox ran off a Raspberry Pi microcomputer and included sensors for monitoring nitrogen oxide, nitrogen dioxide, ozone, and volatile organic compounds (VOCs), as well as wind speed, humidity, and temperature. A logbook provided instructions for using the various part of the kit, and participants were able to upload the data they gathered to the Citizen Sense Kit platform.

Gabrys and Pritchard

The Citizen Sense Kit was used by residents living in a wide range of locations near fracking infrastructure and was also taken up by a local group, Breathe Easy Susquehanna County, which was interested in maintaining constructive dialogue with industry about changes in the environment, particularly in relation to air quality.[16] The kit provided accessible and unobtrusive ways for participants to document pollution events and experiences for three to six months and to observe patterns and relations that emerged from collected data. This approach to environmental sensing was important for a number of reasons. Due to fears of reprisal from neighbors and industry in the sensitive context of air quality monitoring of fracking infrastructure, many participants needed to take part anonymously. The small size of the kit meant that the sensors could be installed and used inconspicuously, such as on porches, in gardens or sheds, or under eaves near homes.

As participating residents lived across the entire area, this resulted in a spatially dense concentration of more than twenty individual monitoring locations, rather than the two or three monitoring points that might be found across rural areas of an entire state. This gave rise to the possibility of identifying localized sources of emissions, which could be read together with the state's air quality data. While data was collected and logged on the Citizen Sense platform, some participants began to notice patterns in their data, particularly in the $PM_{2.5}$ datasets. Using the data together with additional sources of weather data, including wind speed and wind direction data from Weather Underground, participants were able to rule out spikes in their data that were most likely caused by regional sources and focus their energies on pollution events of more than six hours in duration when the wind speed was lower, which would indicate a more local pollution source. The participants who knew each other also formed groups so they could compare one another's data.

The Citizen Sense Kit for monitoring air quality did not just focus on the gathering of numeric data. Photographs, mobile phone videos, YouTube comments, thermal imagery, diaries, and multiple other forms of documentation that might otherwise have seemed like a disparate set of resources all contributed to the making of a "just good enough" collective dataset for the region. Participants were further able to use their local expertise about fracking processes and infrastructure—particularly in relation to compressor stations—to answer questions from regulators, who had little day-to-day experience of living so close to natural gas extraction infrastructure. Another participant set up two monitors at a site opposite a location that was scheduled to be fracked. The unpredictable timing of much fracking activity made it difficult for regulators to monitor a well pad from start to completion. Companies that may have had a permit to drill a well for as long as five years often start fracking without warning. One participant who passed a potential well pad site daily was able to establish a period of monitoring data both before and during the fracking operation. This monitoring, combined with the participant's daily YouTube videos documenting ongoing drilling and fracking, has contributed to a unique set of evidence that can be read alongside more official regulatory monitoring data.

Although in the regulators' view the data generated by the Citizen Sense research project was not comparable to AQI data, it was "just good enough" for the participants to read together with state-collected air quality data and locally collected wind data from Weather Underground. The distribution of devices also contributed to recognizing a regional source of $PM_{2.5}$ in the area. One device on its own probably would not have been "just good enough" to recognize it, but the distribution of multiple devices, maintained by participants on a day-to-day basis over six months, made the data useful for discussion with regulators, since in some cases even regulators and industry are unsure what is being emitted from these sites of concern.

Data that emerged through these techniques also became a useful negotiation tool. It was used to arrange a number of conference calls with regulatory bodies, such as the Centers for Disease Control and Prevention, Agency for Toxic Substances and Disease Registry, Pennsylvania Department of Health, and nonprofit environmental organizations, as well as local political representatives. Responses to the citizen-collected data ran the gamut from outright dismissal to interest. There was just enough evidence to lead one environmental agency to request that local monitoring be undertaken, something Breathe Easy Susquehanna County participants had been campaigning for since the inception of their organization.

Gabrys and Pritchard

Conclusions

Although some citizen sensing projects have worked closely with regulators and scientific disciplines, many others have departed from these practices and instead have used devices in unconventional ways, creating environmental data infrastructures that might be very different both spatially and temporally from those of the regulators. As citizen science and citizen sensing stabilize, there is a call for practices to become more standardized to enhance their legitimacy. Indeed, the EPA cites a need to standardize protocols and datasets, and both the North American Citizen Science Association and the European Citizen Science Association (ECSA) cite a need to establish best-practices guidelines as central to the aims of the organizations.[17] Gatherings such as the ECSA assembly and Citizen Science Center Zurich are coming together specifically to address this problem of standardization.

Much of the ongoing debate by practitioners and organizations that we have observed in citizen science meetings focuses on the importance of developing practices that can be directly comparable to existing regulatory practices. To some extent, there is a gap between the current citizen sensing infrastructure and this vision of comparability. This has led to a drive toward designing devices that create data in similar formats and calibrating devices in reference to regulatory monitoring equipment. In some cases we have observed regulators recommending that citizens monitor only in scenarios that are preapproved by official bodies. Yet in this context the inevitable question is what new possibilities for environmental monitoring and citizen-gathered data might be missed by attending only to the ways in which citizen sensing practices can replicate monitoring practices focused on regulatory compliance.

Citizen-gathered data using next-generation environmental sensing could have multiple uses, and the trajectories of citizen sensing initiatives in making connections from environmental data to action do not need to exclude data that does not conform to regulatory practices or might have a more speculative starting point. Such an approach would imply that any production of data by citizens that does not aim toward regulatory targets and processes could not be useful. We recognize that for some contexts these new infrastructures have proved challenging to both regulators and scientists, whose disciplines and professions have established practices of measuring, monitoring, and accounting for environments. This has often created points of tension and disagreement between regulators and citizens.

But making citizen sensing practices and data conform only to regulatory standards would be to exclude the other creative and political possibilities of what we are calling "just good enough data." To align data practices exclusively with regulatory modes of monitoring might even preclude citizens from participation in citizen sensing altogether. For instance, to be comparable to the state DEP and EPA air quality data in the context of monitoring for $PM_{2.5}$, citizen data would need to be collected by officially trained personnel on approved equipment. Furthermore, this monitoring would have to take place at the same location, height, and position at which regulatory monitoring is already situated. In the context of the AQI $PM_{2.5}$ monitoring, citizen monitoring would also have to be done over a time span of three years. One might argue that as the data analysis process involves many decisions, data would have to be analyzed (including averaged and smoothed) using the same software and algorithms as in the state and federal processes. In this scenario, citizen sensing as a practice would become completely redundant, as it would have to replicate the monitoring performed by governmental agencies and experts, rather than opening up opportunities for monitoring to be undertaken by a wider range of participants, in varied locations, over different timescales, and in response to distinct events.

Instead, we suggest that "just good enough data"—while not ignoring the important issues of accurate instrumentation, calibration, and measurement, along with robust monitoring practices—might allow for more expansive uses of citizen sensing technologies and data. It might do so while still opening up a dialogue on environmental change between citizens and regulators. With this proposal, we are not regressing to earlier conceptions of public science, where the collection of data is a cursory one oriented toward raising public awareness. On the contrary, we suggest that "just good enough data" is a practice and emerging infrastructure that creates a shared space for discussion

Gabrys and Pritchard

where citizens can exchange community understandings of pollution events with regulators. Citizen-produced datasets are often declared to be inaccurate because of the devices used, illegitimate because of the protocols followed, and unscientific because of perceived community bias (such as citizens monitoring to create deliberate evidence for pollution events). However, we have shown that citizen sensing is also an entry point for testing claims about the ease of participation that next-generation environmental sensors are meant to offer, as well as for developing expanded aspects of monitoring, data collection, and environmental politics that might allow communities to engage more readily with environmental problems.

Acknowledgments

The research leading to these results has received funding from the European Research Council under the European Union's Seventh Framework Programme (FP/2007–2013) / ERC Grant Agreement n. 313347, "Citizen Sensing and Environmental Practice: Assessing Participatory Engagements with Environments through Sensor Technologies." Thanks are due to participating residents in Pennsylvania, including Frank Finan, Rebecca Roter, Meryl Solar, Vera Scroggins, Chuck and Janis Winschuh, Paul Karpich, Barbara Clifford, John Hotvedt, Barbara Scott, Audrey Gozdiskowski, Alex Lotorto, and anonymous participants, as well as previous Citizen Sense researchers, including Nerea Calvillo, Tom Keene, and Nick Shapiro, and consultants including Kelly Finan (illustration), Benjamin Barratt (atmospheric science), Raphael Faeh (web design), Catherine Pancake (video documentation), and Lau Thiam Kok (data architecture). We are also grateful to the CREATE Lab at Carnegie Mellon University for loaning the Speck devices for use in this study.

Notes

1. An extensive set of literature exists on environmental data, which the essays in this book discuss and extend. For additional references on environmental data in relation to institutional infrastructures as discussed here, see Bowker, "Biodiversity Datadiversity"; Edwards, *Vast Machine*; Fortun, "Biopolitics"; Gabrys, "Practicing, Materializing and Contesting"; Lippert, "Environment as Datascape"; Nadim, "Blind Regards"; Turn-

hout, Neves, and de Lijster, "'Measurementality' in Biodiversity Governance." For a discussion of the vagaries of sensor data in particular, see Nafus, "Stuck Data, Dead Data."

2. For a more extensive discussion on the uses of environmental data and citizen data for social and environmental justice, see Gabrys, "Data Citizens"; Walker et al, "Practicing Environmental Data Justice."

3. Burke et al., "Participatory Sensing"; Cuff, Hansen, and Kang, "Urban Sensing"; Elwood, "Volunteered Geographic Information." Michael F. Goodchild has discussed the concept of "citizens as sensors" in the context of volunteered geographic information; however, his proposal does not encompass actual sensor devices as usually understood within computational and environmental sensing practices. See Goodchild, "Citizens as Sensors."

4. Aoki et al., "Common Sense"; Maisonneuve et al., "Citizen Noise Pollution Monitoring"; Paulos, Honicky, and Hooker, "Citizen Science."

5. Irwin, *Citizen Science*; Jasanoff, "Technologies of Humility."

6. EPA, *Draft Roadmap*.

7. Gabrys, Pritchard, and Barratt, "Just Good Enough Data."

8. Citizen Sense, http://citizensense.net.

9. Gabrys, *Program Earth*.

10. StateImpact, "Marcellus Shale."

11. Fractracker Alliance, last updated November 2021, https://www.fractracker.org/map/us/pennsylvania/pa-shale-viewer.

12. MarcellusGas.org, last updated February 5, 2022, https://web.archive.org/web/20220121233306/https://www.marcellusgas.org/.

13. As a result of the citizen monitoring efforts, the DEP expanded its air quality monitoring infrastructure in the area, with monitors installed and in place as of November 2018. This more recent development is discussed further in Gabrys, *Citizens of Worlds*. See also Hurdle, "PA Expands Particulate Monitoring."

14. While there is not space in this chapter to unpack the particular collaborative methods developed in the Citizen Sense project, we drew in part on earlier comparable studies on sensors and engagement. For example, see DiSalvo et al., "Towards a Public Rhetoric." For a more in-depth discussion of the methods developed in the Citizen Sense project, see Gabrys, *How to Do Things with Sensors*.

15. See Moore, "Air Impacts."

16. Breathe Easy Susquehanna County, https://www.facebook.com/BreatheEasySusq.

17. EPA, *Citizen Science and Crowdsourcing*; Citizen Science Association, "Vision, Mission, Goals"; ESCA, "Ten Principles of Citizen Science."

Bibliography

Aoki, Paul M., R. J. Honicky, Alan Mainwaring, Chris Myers, Eric Paulos, Sushmita Subramanian, and Allison Woodruff. "Common Sense: Mobile Environmental Sensing Platforms to Support Community Action and Citizen Science." In *Adjunct Proceedings UbiComp 2008*, 59–60. Seoul: ACM.

Bowker, Geoffrey C. "Biodiversity Datadiversity." *Social Studies of Science* 30, no. 5 (2000): 643–83.

Burke, Jeff, Deborah Estrin, Mark Hansen, Andrew Parker, Nithya Ramanathan, Sasank Reddy, and Mani B. Srivanstava. "Participatory Sensing." In *Proceedings of the World Sensor Web Workshop*. Boulder CO: ACM SENSYS, 2006.

Citizen Science Association. "Vision, Mission, Goals." 2014. http://citizenscienceassociation .org/overview/goals.

Cuff, Dana, Mark Hansen, and Jerry Kang. "Urban Sensing: Out of the Woods." *Communications of the Association for Computing Machinery* 51, no. 3 (2008): 24–33.

DiSalvo, Carl, Marti Louw, David Holstius, Illah Nourbakhsh, and Ayca Akin. "Towards a Public Rhetoric through Participatory Design: Critical Engagements and Creative Expression in the Neighborhood Networks Project." *Design Issues* 28, no. 3 (2012): 48–61.

Edwards, Paul N. *A Vast Machine: Computer Models, Climate Data, and the Politics of Global Warming*. Cambridge MA: MIT Press, 2010.

Elwood, Sarah. "Volunteered Geographic Information: Future Research Directions Motivated by Critical, Participatory, and Feminist GIS." *GeoJournal* 72, nos. 3–4 (2008): 173–83.

EPA (U.S. Environmental Protection Agency). *Citizen Science and Crowdsourcing: Creative Approaches to Environmental Protection*. Draft report, September 2015. http://www.epa .gov/sites/production/files/2015-09/documents/nacept_background_material_2.pdf.

———. *Draft Roadmap for Next Generation Air Monitoring*. Draft report, March 8, 2013. https://www.epa.gov/sites/production/files/2014-09/documents/roadmap -20130308.pdf.

ESCA (European Citizen Science Association). "Ten Principles of Citizen Science." September 2015. https://ecsa.citizen-science.net/wp-content/uploads/2020/02/ecsa _ten_principles_of_citizen_science.pdf.

Fortun, Kim. "Biopolitics and the Informating of Environmentalism." In *Lively Capital: Biotechnologies, Ethics, and Governance in Global Markets*, edited by Kaushik Sunder Rajan, 306–26. Durham NC: Duke University Press.

Gabrys, Jennifer. *Citizens of Worlds*. Minneapolis: University of Minnesota Press, forthcoming.

———. "Data Citizens: How to Reinvent Rights." In *Data Politics: Worlds, Subjects, Rights*, edited by Didier Bigo, Engin Isin, and Evelyn Ruppert, 248–66. New York: Routledge Studies in International Political Sociology, 2019.

———. *How to Do Things with Sensors*. Minneapolis: University of Minnesota Press, 2019.

———. "Practicing, Materializing and Contesting Environmental Data." *Big Data & Society* 3, no. 2 (2016): 1–7.

———. *Program Earth: Environmental Sensing Technology and the Making of a Computational Planet*. Minneapolis: University of Minnesota Press, 2016.

Gabrys, Jennifer, Helen Pritchard, and Benjamin Barratt. "Just Good Enough Data: Figuring Data Citizenships through Air Pollution Sensing and Data Stories." *Big Data & Society* 3, no. 2 (December 2016): 205395171667967.

Goodchild, Michael F. "Citizens as Sensors: The World of Volunteered Geography." *GeoJournal* 69 (2007): 211–21.

Hurdle, Jon. "PA Expands Particulate Monitoring as Federal Study Finds High Level in One Location." StateImpact Pennsylvania, May 4, 2016. https://stateimpact.npr.org/pennsylvania/2016/05/05/pa-expands-particulate-monitoring-as-federal-study-finds-high-level-in-one-location/.

Irwin, Alan. *Citizen Science: A Study of People, Expertise and Sustainable Development.* London: Routledge, 1995.

Jasanoff, Sheila. "Technologies of Humility: Citizen Participation in Governing Science." *Minerva* 41 (2003): 223–44.

Lippert, Ingmar. "Environment as Datascape: Enacting Emission Realities in Corporate Carbon Accounting." *Geoforum* 66 (2015): 126–35.

Maisonneuve, Nicholas, Matthias Stevens, Maria E. Niessen, Peter Hanappe, and Luc Steels. "Citizen Noise Pollution Monitoring." In *Proceedings of the 10th International Digital Government Research Conference* (May 17–21, 2009), 96–103. Puebla, Mexico.

Moore, Chris. "Air Impacts of Gas Shale Extraction and Distribution." Presented at the Workshop on Risks of Unconventional Shale Gas Development, Washington DC, May 30–31, 2013. http://sites.nationalacademies.org/cs/groups/dbassesite/documents/webpage/dbasse_083402.pdf.

Nadim, Tahani. "Blind Regards: Troubling Data and Their Sentinels." *Big Data & Society* 3, no. 2 (2016): 1–6.

Nafus, Dawn. "Stuck Data, Dead Data, and Disloyal Data: The Stops and Starts in Making Numbers into Social Practices." *Distinktion: Journal of Social Theory* 15, no. 2 (2015): 208–22.

Paulos Eric, R. J. Honicky, and Ben Hooker. "Citizen Science: Enabling Participatory Urbanism." In *Handbook of Research on Urban Informatics: The Practice and Promise of the Real-Time City*, edited by Marcus Foth, 414–36. Hershey PA: Information Science Reference, 2009.

StateImpact. "The Marcellus Shale, Explained." StateImpact Pennsylvania, February 2020. http://stateimpact.npr.org/pennsylvania/tag/marcellus-shale.

Turnhout, Esther, Katja Neves, and Elisa de Lijster. "'Measurementality' in Biodiversity Governance: Knowledge, Transparency, and the Intergovernmental Science-Policy Platform on Biodiversity and Ecosystem Services (IPBES)." *Environment and Planning A* 46, no. 3 (2014): 581–97.

Walker, Dawn, Eric Nost, Aaron Lemelin, Rebecca Lave, and Lindsey Dillon. "Practicing Environmental Data Justice: From DataRescue to Data Together." *Geo: Geography and Environment* 5, no. 2 (July–December 2018): 1–14.

SIX

Collaborative Modeling as Sociotechnical
Data Infrastructure in Rural Zimbabwe

M. V. Eitzel, Jon Solera, K. B. Wilson, Abraham Mawere Ndlovu, Emmanuel Mhike Hove, Daniel Ndlovu, Abraham Changarara, Alice Ndlovu, Kleber Neves, Adnomore Chirindira, Oluwasola E. Omoju, Aaron C. Fisher, and André Veski

The idea that we can use ever-larger datasets and more complex models to combat perils of the Anthropocene, such as mass extinction or climate change, is increasingly prevalent within environmental science.[1] The assumption is that aggregating big data and using machine learning to analyze this data will lead scientists to better understandings of nature and society's relationships with it.[2] However, even when these modeling techniques acknowledge the inherent inseparability of nature and society, they can reinforce faulty assumptions and unjust practices rather than contribute new knowledge.

Restricting our research to only those processes about which we collect large volumes of quantitative data limits what decision-makers manage for.[3] In addition, uncritically importing the predictive algorithmic approaches of industry into universities, governments, and nongovernmental organizations, an increasingly common practice in which only the powerful have the ability to correct erroneous model assumptions or results, has serious consequences for democracy.[4] Often the opacity of this modeling infrastructure is penetrable only by relatively wealthy and formally educated people, while at the same time having disproportionate and negative impacts on the most vulnerable.[5] Big data–focused modeling practices can therefore harm the environment (through limited understanding) and democracy (through limited access) and can narrow our view of human relationships with and within nature.

Collaborative modeling, including participatory mapping, is an alternative approach to using automated big-data approaches to know and govern nature.[6] By thoughtfully bringing together outsider knowledge with community-based knowledge—and contextualizing or situating those knowledges—collaborative modeling can drive "better accounts of the world," recognizing the partial contribution of each to understanding and enabling "connections and unexpected openings."[7] It can support and feature local and Indigenous knowledge within socioecological systems and highlight how the relationship between outsiders and communities affects knowledge production.[8] Truly participatory strategies that involve or are led by communities at all stages of scientific practice have the potential to leverage outside partners' authority while at the same time reconfiguring the notion of expertise itself.[9] This reconfiguration is crucial in environmental management and decision-making, where whose knowledge counts as legitimate often determines the "winners" and "losers" in a particular situation.[10] Given colonial legacies, reconfiguring environmental expertise is particularly important in Indigenous resource management.

Practiced with a critical eye, collaborative modeling approaches that focus on community-driven questions and iterative model development can prove an antidote to the risks of automated big-data techniques. Community-driven questions can ensure the relevance of research and justify the contribution of data and labor from Indigenous groups. Iterating model design between communities and modelers—a feedback step lacking in many algorithmic, big data–oriented processes—enables changes to models or data analysis that can lead to better understanding of the system under study, as well as more just outcomes. Realizing the potential of collaborative modeling therefore requires that communities direct the framing, data input, and interpretation of the modeling: a "modeling of the oppressed" approach (following Paulo Freire's "pedagogy of the oppressed").[11] Decolonizing knowledge production becomes more possible if modeling processes also honor the potential need for action alongside observation and support the long-term process necessary for building intentional relationships between outside researchers and Indigenous communities.[12] To reimagine how modeling and data analysis lead to knowledge about socioecological systems—leaving room for multiple knowledges to co-contribute—we need data infrastructures that explicitly attend to the social and relational aspects of the modeling process, not just its technical aspects.

Eitzel et al.

Data becomes infrastructure in the context of how people use it, especially when it becomes structural, supportive, and taken as a given.[13] For such infrastructures to be reliable, useful to local and Indigenous communities, and potentially transformative in the face of the pressure for automatic application of big-data practices, modelers must prioritize investments in the social components of their data and modeling. On the technical side, modeling infrastructure includes software, hardware, and data. But social components of these infrastructures include relationships, attitudes, organizations, and skills—anything that the team must rely on to work invisibly in the background.[14] While relationships are foundational to collaborative modeling, they require conscious construction and maintenance. This framing of infrastructure as both technical and social raises the question of how these components can mutually reinforce each other, becoming sociotechnical infrastructures.[15] Successful collaborative modeling benefits from an awareness of how software and other technical tools facilitate relationship building among all members of a research team and how such relationships in turn facilitate further technical work.

In this chapter, we take an autoethnographic approach to investigate sociotechnical collaborative modeling infrastructure.[16] We analyze our own collaborative modeling process between outside analysts and the Muonde Trust, a registered Zimbabwean nongovernmental organization dedicated to supporting Indigenous innovation in Mazvihwa Communal Area, Midlands Province. We use a situational analysis to examine the social and technical infrastructures supporting this modeling process and to ask how the infrastructures support each other—that is, to what degree they form a sociotechnical infrastructure.[17]

Our analysis of the modeling process is based on participant observation during the creation and application of the Zimbabwe Agro-Pastoral Management Model, deriving chiefly from field notes written by several of the chapter authors during modeling processes and workshops. Other sources include autoethnographic notes by the first author of this chapter, M. V. Eitzel; email threads and conversations via WhatsApp (a cell phone text messaging service) with other coauthors; and all the versions of model code that were created in the course of the project. Eitzel analyzed these materials using a grounded theory approach, creating memos during model development and discussion.[18] The memos and notes were then iteratively coded to generate themes.[19] These themes became the basis of interview questions sent by email and WhatsApp, as well as group interview questions for later workshops.

Eitzel conducted a situational analysis following each community-based workshop, creating and then revising an ordered/working version of a situational map, a systematic list of concepts, individuals, collectives, and nonhuman actors associated with the issue or situation.[20] We selected all elements from the situational map relating to infrastructure to analyze in this chapter. Eitzel wrote a thick description of the modeling process, which was later distilled into the description that appears in this chapter.[21] Finally, drafts of the analysis and framing of this chapter were shared with all coauthors to allow for additional themes and ideas to emerge. In the next section we briefly describe the Muonde Trust's community research team; the pressing stewardship questions in Mazvihwa Communal Area regarding land use, management intervention, and climate change; and our collaborative modeling process. Subsequent sections outline the technical infrastructures present in the process, the social infrastructures, and the ways in which they mutually reinforced each other to become sociotechnical infrastructures.

Note that we refer to participants in our collaborative process in several different ways. We often specifically refer to Eitzel, who did the majority of the analysis and writing of this chapter, but we also follow principles of participatory action research and refer to outside researchers and community researchers to distinguish who is embedded in the system under study.[22] We often refer to chapter coauthor K. B. Wilson explicitly because he bridges this gap as an extremely embedded outsider.[23] At some points, we refer to participants based on their role or skills (for example, programmers) or specific contexts (Santa Fe Institute students, local leaders, or Muonde researchers). We prefer flexible terminology to acknowledge the situations of various contributors while not unnecessarily emphasizing their differences.

The Muonde Trust: Long-Term Community-Based Participatory
Research and Collaborative Modeling

Mazvihwa Communal Area is categorized in the driest and lowest-potential agroecological zone of Zimbabwe. Historically, the colonial government had reserved most of the country's better agricultural land for itself and for commercial interests, increasingly restricting Indigenous Zimbabweans to tribal reservations such as Mazvihwa. People initially survived there by intensively cultivating small natural wetlands coupled with pastoralism, as well as hunting and gathering over large areas. However, as part of the colonial government's efforts to modernize agriculture and moderate land degradation in African areas that had become densely populated, wetland agriculture was banned, and settlement patterns were shifted to previously wooded watersheds.[24] Even after independence in 1980 the Zimbabwean government has struggled to find a solution to land management issues, both within communities like Mazvihwa and at the national level, where the postcolonial redistribution of lands has been replete with controversy.

The Muonde Trust formed based on an initial collaboration between K. B. Wilson, Billy B. Mukamuri, and Abraham Mawere Ndlovu in the mid-1980s, working to study the resilience of households making use of the different soil types found in Mazvihwa and developing a suite of action research–based community natural resource management and agricultural efforts. Following this initial research, they have continued to collect data to answer questions of interest to the community living in Mazvihwa, over time building up a research team of about thirty people of various ages. The team has been composed largely of women farmers from several villages in northeastern Mazvihwa, with representatives of many of the local clans, including those from the traditional ruling clan (Hove), as well as people who have held local elected offices. Over the thirty years following Wilson and Mawere Ndlovu's initial collaboration, Muonde's community research team measured and recorded a variety of information about the farming and livestock management practices, ranging from crop yields and livestock counts to narratives about historical management situations and information on locally adapted small grain varieties. Muonde also maintains an openness to outsiders who can bring knowledge and skills on topics of interest to them.[25]

Eitzel was just such an outsider, encouraged by Wilson to work with Muonde after seeing Wilson and Mawere Ndlovu present the summary of the trust's research in a graduate seminar. We discussed the potential of modeling to help synthesize the community's data and address its concerns regarding land management. More specifically, the Muonde research team's key question was how much land to allocate to agricultural production and how much to leave as woodland grazing area for livestock. Additional considerations included investigating interactions between crop and livestock management interventions, both novel and traditional, and possible larger year-to-year variation in rainfall due to climate change.

The livestock, farming, and woodland management practices the Muonde team had recorded and wanted to study influenced each other in an intricate set of feedbacks that Eitzel and Wilson set out to represent in an agent-based model (ABM). ABMs are simulation-based models in which modeled agents make decisions dynamically based on programmer-specified rules, typically interacting with other agents and with the modeled landscape. Together with students from the Santa Fe Institute (SFI) Complex Systems Summer School (K. Neves, A. Veski, and O. Omoju), we integrated information from the Muonde team, which varied in its level of quantification, into ABM rules in the programming language NetLogo for how the crops, livestock, and woodland grazing area interacted with each other and changed over time based on rainfall.[26] We eventually published a description of the model rules, parameters, and the Muonde data they were based on, along with the results of a wide range of simulations.[27]

In addition to consistently checking in with the Muonde team and larger Mazvihwa community to ensure validity of the model, we also checked the results against their own data.[28] We were advised by ABM experts to be conservative with the results of these kinds of models, which aligned with our approach toward the community: to emphasize that this model did not replace their own knowledge of their system and was meant to generate discussion.[29] We conducted workshops with small groups (fig. 8) where the Muonde research team members

Eitzel et al.

ran the model and discussed both it and the real system. These workshops were followed by a large, whole-group workshop including all thirty members of the team, in which we shared and synthesized the team's experiences and reflections about the model. The Muonde team later conducted community workshops in which the team shared the model with local leaders and other members of the community, reporting back via WhatsApp on the results of these workshops to Eitzel.

Technical Infrastructures: Programming Languages, Communication Technologies, and Data

Technical infrastructures that supported the modeling effort included the type of model produced, communication technologies and international travel that facilitated international teamwork, and Muonde's rich data archive. Agent-based modeling, with its focus on mechanisms and individual-level granularity, lends itself to collaborative modeling by providing a relatively easy way to integrate local and Indigenous peoples' detailed knowledge about their systems into a computer modeling framework (for example, in rules governing agent behavior). The choice of the ABM for this project was partly for convenience because of the topics of the SFI Summer School and the interests of the students, but it emerged as a good choice for integrating the wide variety of qualitative and quantitative information we had about the system.[30] NetLogo was also very useful in rapidly developing a user interface that was relatively easy for the community to relate to and interact with (this was no accident, as it was designed initially as a teaching tool).[31] We noted this ease relative to the community's previous experience learning QGIS (a geographic information system application) for digital mapping tasks; QGIS interface is much more complex and abstract than the interface we developed in NetLogo.[32] A larger number of community members picked up NetLogo's interface and expressed more confidence in its use than with QGIS.

NetLogo later proved to be an unfortunate choice, however, when we tried to run many combinations of parameter choices for debugging and sensitivity analysis, because the Java-based code runs slowly and can only partially be run in parallel (that is, it was difficult to efficiently run many models at the same time). The Python programming language would have worked better for the software and scientific aspects of the model, but creating a graphical user interface in Python is more difficult than in NetLogo. Therefore, NetLogo prioritized the usefulness of the model for the community rather than its scientific validation.

In addition to the programming infrastructure used to create the model, we made heavy use of communication technologies. We relied on WhatsApp for text messaging with the Muonde research team and on Skype for videoconference calling between other team members all across the world. We also made use of Dropbox as a file-sharing and version-control method when we were actively collectively developing the model code. One programmer suggested that we use GitHub as a version control system, but we preferred Dropbox because GitHub does not allow private archives without a paid subscription, and the community had not yet given their approval to release the model code to the world. We also used email clients and mailing lists to keep outside researchers apprised of progress, ask interview questions, and send the current model versions to the Muonde research team members. Coordinating an international team was extremely dependent on these communication technologies.

Equally important was the international travel of the summer school students to Santa Fe, outside researchers to Mazvihwa, and Muonde researchers to California. Wilson's initial visits to Mazvihwa in the 1980s were key to building Muonde's social and technical infrastructures, and Eitzel's involvement would have been unlikely if Mawere Ndlovu had not traveled to the United States to synthesize and present the Muonde data archive with Wilson. Particularly key international travel occurred during initial model development at the summer school, when the students arrived in Santa Fe and Mawere Ndlovu was in California and could advise them via Skype (along with Wilson) as they made the first versions of the model and developed lists of features to implement in the future. Finally, Eitzel's travel to work with the Muonde team on presenting drafts of the model in person was critical to both teaching how the model worked and getting feedback to improve it.

The data infrastructure we drew on includes access via university subscriptions to a wealth of agroecological research from all over Africa to

8. Social, technical, and sociotechnical data infrastructures. Using technical components such as laptops, electricity, programming languages, and communication technologies, our team was able to create a model in a form that the Muonde Trust research team could give feedback on and then later share with local leaders and community members. The model provided a project for outside researchers to work on with the community researchers, feeding into trusting relationships (e.g., between outsider M. V. Eitzel, seated at right, and translator Abraham Changarara, standing second from right), which supported further technical development. Community feedback, enabled by Muonde's leadership (e.g., executive director Abraham Mawere Ndlovu on far left) and the research team, improved the model, and the model supported the research team's advocacy for land-use decisions that could preserve important cultural resources, mutually reinforcing each other and forming a sociotechnical infrastructure. Photo by Jon Solera. Used with permission.

find support for model parameters, which was also facilitated by consistent internet access and electricity. (Utilities are not always available to the community researchers in Zimbabwe.) The archive of Muonde's data was also extremely important as technical infrastructure, as was access to research team members to answer questions about what to include in the model and which emergent behaviors in the model were realistic versus which were mistakes to be fixed. However, one weakness in the data infrastructure of this project is the reliance on key individuals, especially Wilson, for access to data (including important metadata). He has maintained the archive for many years, and though the physical records have been digitized, much of the contextual information—including how to interpret potential meanings of quantitative findings—is stored only in his mind, even when parts of his knowledge have been analyzed and written up for others. In addition, there is no single summary of all the data available, and data collected on the ground is not always sent to outsiders to curate. Until recently, the Muonde Trust lacked a central location to store data that has been gathered, leading to various datasets being frequently misplaced. With the recent construction of Muonde's office and headquarters, it is more likely that data curation in Mazvihwa will improve.

Social Infrastructures: Affective Engagement, Friendships, Facilitation Skills, and Programming Skills

Social infrastructures that were key to the success of the modeling included emotional engagement, relationships built and maintained between various team members, and skill development. In terms of affect, it was key for outside researchers to maintain an attitude of humility, patience, and flexibility to allow community members to learn the model at their own pace and ensure that their place- and experience-based knowledge was not displaced by model-based knowledge. Community members benefited from confidence building around the model and computer use skills, bolstered by Muonde's long-standing research ethic and recent work on participatory mapping.[33] Affect also structured the relationship between people and the model: programmers were genuinely excited when watching the simulated cows move according to the rules they had established and concerned when the cow population crashed or the woodland became extremely denuded. The community similarly connected to the simulated cows, but with the added dimension that they remembered the real die-offs of their own cattle. Emotional connection with model processes helped programmers and community members alike engage and identify with the model and its outcomes, and it enabled community members to tap more effectively into strong memories.

These emotional connections were facilitated by the long-standing relationships between Muonde research team members and outside researchers, starting with Wilson and Mawere Ndlovu, and more recently with Eitzel and younger members of the team, including E. Mhike Hove, A. Ndlovu, A. Changarara, and D. Ndlovu. Friendships and research partnerships, developed over time via continued connection and sharing of life events, as well as professional and community news, sustained these team members' abilities to work together on a much more technical task like model creation and presentation. Similarly, Muonde has long-standing relationships with many people in the Mazvihwa community, including local leaders, and the ultimate outcomes of the model regarding land use change would not have been possible without the maintenance and additional strengthening of these connections. For outside researchers, friendships begun at s f i and elsewhere facilitated teamwork at a scale that several people had not previously experienced. One researcher explains, "I think the surprising part was the smoothness of cooperation within the team," despite its large size. As different team members took up the programming tasks over time, it became clear that they all left their own sense of humor in the code's variable names and comments. Even in the final versions of the model available online, core parts of the code are intact from the original teamwork at s f i, a detail that was important to some of the modelers in order to respect and acknowledge the work of the initial programmers.

Finally, participants in this project learned and practiced a variety of skills. Several outside researchers learned to program in NetLogo; Eitzel learned high-performance computing techniques and a variety of programming best practices, such as profiling to check which parts of the code were running slowly and writing unit tests to ensure the smallest pieces of the code would do what they were intended to do. Muonde team members learned how to use the model and present it to each other and to other community members. Facilitation skills were key as well, including "coordination abilities, a minimum knowledge of the socio-ecological processes in play, a certain ease with computer tools and an undoubted ability for dialogue and exchanging information."[34] These were skills held and reinforced by Eitzel on this project, and several other outsiders learned some community collaboration skills as well.[35] In addition, though the outside researchers were all from different academic and engineering disciplines, they were skilled in interdisciplinary collaboration, which facilitated the integration of a variety of methods and perspectives into the modeling process, as well as the ability to communicate with each other about these methods and perspectives.

Sociotechnical Infrastructures and Collaborative Modeling: Land Use Planning in Mazvihwa

The infrastructure supporting our modeling processes was decidedly sociotechnical, in the sense that the technical and social infrastructures mutually reinforced each other. The friendships and collaborations among team members would have been impossible without the communication technologies, especially WhatsApp and Skype. Though the ABM relied heavily on the data that the Muonde Trust had gathered over the years, as archived by Wilson, the data could be collected only in the context of Wilson's continuing relationships with the Muonde researchers and their attitudes of confidence toward research as a practice for addressing their community's problems. The work would not have been possible without resources from the U.S. National Science Foundation, but that grant application was successful based on the strength of Wilson's connection with the Muonde Trust and his mentorship of Eitzel as the principal investigator of the grant, as well as the strength of her technical modeling skills. Creating the ABM was a direct consequence of the requirement to complete a project associated with the SFI Summer School, but NetLogo by itself would not have been useful without other students willing to learn how to program in it while also learning content knowledge about the agroecological system—an additional step facilitated both by Skype and by collaborative relationships with Wilson and Mawere Ndlovu.

Muonde's social and technological infrastructures have also mutually supported each other over the years. Their relationships with community members and local leaders facilitated access to people and resources for measuring the aspects of their agropastoral system that ultimately fed into their data archive, which made our model possible. Wilson's ability to get outside collaborators to travel to Zimbabwe to work with the community team was based on his relationships with those collab-

Eitzel et al.

orators and with the Muonde team, but international travel was a pre-requisite for the outside collaborators to form relationships with the Muonde team. The previous year, Eitzel and Jon Solera had already traveled to Mazvihwa, allowing them to build relationships and work with the community team on participatory digital mapping projects, so the prior travel and relationships were key in supporting the further technical work of this project's modeling process.[36]

Muonde's relationships with Mazvihwa community members and authorities were also essential in the application of the model to generate action regarding the original question of interest. The Muonde research team demonstrated the model to local leaders and discussed the issues of dwindling woodland grazing area with them. By convening the community and decision-makers, using the model as a focus, Muonde researchers were able to propose solutions to the problem and are now piloting a new policy in which fallow fields owned by absentee landowners can be cultivated by village residents, rather than creating new crop fields in the woodland.[37] Muonde also has an important role in helping develop a biocultural restoration protocol for the community's *rambotemwa* (sacred forests), and has generated much interest in reforesting and afforesting the degraded parts of these forests. The model is unquestionably a technical component of Muonde's ability to generate this outcome, but it was the combination of the model with the relationships with local leaders, and the Muonde team members' confidence in their ability to do research, that enabled on-the-ground change.

Conclusions

Unreflective use of big-data approaches to assess socioecological systems is likely to recapitulate unjust colonial relationships between people and their environments, between people and models, and between different groups of people. Collaborative modeling promises to transform these relationships, particularly when projects are community led. But critical review is still warranted. We first highlight the successes of our modeling process: it resulted in a tool that the community could use to take action on a pressing need regarding land-use planning, and it foregrounded the building and sustaining of long-term relationships—both features necessary for moving toward transformative practice.[38] Our technical and social infrastructures mutually supported each other and together facilitated the collaborative process. Our model was developed with and for the community, using the community's data to address its questions and taking iterative development steps that allowed outside modelers to ensure that the model reflected the realities and priorities of the community.[39] These facets of our work depart from any modeling that had previously been applied to Mazvihwa and represent a move toward decolonizing knowledge production.

Of course, a critical approach to collaborative modeling requires us to investigate how our process may still perpetuate colonial knowledge regimes. We note that the community did not participate in the choice of the modeling tool itself, nor did community members learn to program the computers themselves. The concept of an ABM and the programming language NetLogo originated outside the community and may carry with them underlying colonial assumptions. Computers were first brought to Mazvihwa in 2013, however, so expecting the Muonde team to construct a program like our ABM—or to develop an alternative—would have been unrealistic. In the future we can aspire to a process in which the community is involved in all modeling choices and takes a hand in the programming, if desired. In addition, the majority of the analytical work underlying this chapter was done by Eitzel, an outsider. Though the community has given feedback on drafts of this chapter, future work could pursue additional infrastructure needed to better facilitate coanalysis and cowriting.

The community is firm in its assertion that the model is its model and will continue to help the community in its land use management and planning. As of 2019 multiple new homesteads were created in the fallow crop fields, rather than being placed in the woodland grazing area—a great success for Muonde's research team in building a more sustainable community. So though there are always limits on just how emancipatory a given modeling project can be, the addition of this novel technical tool to Muonde's already rich sociotechnical infrastructure has indeed had a positive impact. We hope that this case study continues to support the trend toward community-led, critical community-based modeling. The larger project of decolonizing knowledge and supporting more just modeling practices requires being watchful of the places where well-meaning outsiders continue to lead unintentionally through their technical infrastructures. We hope that recognition of the social and sociotechnical aspects of the modeling process will increasingly enter into training for outside researchers aiming to assist communities with their modeling needs and that communities will increasingly generate their own ideas for what modeling should look like.

Acknowledgments

We acknowledge all the members of the Muonde Trust who have worked on the research over the years, and especially the participants in our modeling workshops: Handsome Madyakuseni, Austen Mugiya, Tatenda Simbini Moyo, Britain Hove, Nehemiah Hove, Khaniziwe Chakavanda, Simon Ndhlovu, Sikhangezile Madzore, Innocent Ndlovu, Blessed Chikunya, Maria Fundu, Lucia Dube, Guilter Shumba, Ndakaziva Hove, Sarah Tobaiwa, Moses Ndhlovu, Adnomore Chirindira, Oliver Chikamba, Cephas Ndhlovu, Esther Banda, Egness Masocha, Abraham Ndhlovu, Princess Moyo, Godknows Chinguo, Nenero Hove, Hosea Ndlovu, Valising Mutombo, Beulah Ngwenya, Ruth Munhundagwa, Vonai Ngwenya, Nyengeterai Ngandu, Saori Ogura, and Alejandra Cano. Katherine Weatherford Darling, Jenny Reardon, Lizzy Hare, and Andrew Mathews gave valuable advice on qualitative analysis, workshop structure, justice framing, and interview questions. We thank the Santa Fe Institute for hosting the Complex Systems Summer School 2015, for which the model was initially developed. Stephen Guerin, Andrew Berdahl, Joshua Epstein, Isaac Ullah, Matthew Potts, and Juan Carlos Castilla provided advice on agent-based models. Eric Nost and Jenny Goldstein gave absolutely essential recommendations on framing and theoretical background for this chapter. This work was supported by the U.S. National Science Foundation (1415130).

Notes

1. Salmond, Tadaki, and Dickson, "Can Big Data Tame?"
2. Hampton et al., "Big Data."

3. Taylor, *Unruly Complexity*.
4. Brown, *Undoing the Demos*.
5. Eubanks, *Automating Inequality*.
6. Étienne, *Companion Modelling*.
7. Haraway, "Situated Knowledges," 590.
8. Robbins, "Beyond Ground Truth."
9. Lave, "Future of Environmental Expertise."
10. Goldman, Nadasdy, and Turner, *Knowing Nature*.
11. Freire, *Pedagogy of the Oppressed*.
12. Coombes, Johnson, and Howitt, "Indigenous Geographies III."
13. Star and Ruhleder, "Steps toward an Ecology."
14. Star and Ruhleder, "Steps toward an Ecology."
15. Edwards, *Vast Machine*.
16. Ellis, Adams, and Bochner, "Autoethnography."
17. Clarke, *Situational Analysis*.
18. Charmaz, *Constructing Grounded Theory*.
19. Cope, "Coding Qualitative Data."
20. Clarke, *Situational Analysis*.
21. Merriam and Tisdell, *Qualitative Research*.
22. Herr and Anderson, *Action Research Dissertation*.
23. Merriam and Tisdell, *Qualitative Research*.
24. Wilson, "Water Used to Be Scattered."
25. Eitzel et al., "Sustainable Development."
26. Wilensky, *NetLogo*.
27. Eitzel et al., "Using Mixed Methods."
28. Eitzel et al., "Using Mixed Methods."
29. Agar, "My Kingdom."
30. Voinov et al., "Tools and Methods."
31. Wilensky and Stroup, "Learning through Participatory Simulations."
32. Eitzel et al., "Sustainable Development."
33. Eitzel et al., "Sustainable Development."
34. Étienne, *Companion Modelling*, 39.
35. Eitzel et al., "Assessing the Potential."
36. Eitzel et al., "Sustainable Development."
37. Eitzel et al., "Assessing the Potential."
38. Coombes, Johnson, and Howitt, "Indigenous Geographies III."
39. Eitzel et al., "Indigenous Climate Adaptation Sovereignty."

Bibliography

Agar, Michael. "My Kingdom for a Function: Modeling Misadventures of the Innu-
    merate." *Journal of Artificial Societies and Social Simulations* 6, no. 3 (Spring 2003).
Brown, Wendy. *Undoing the Demos: Neoliberalism's Stealth Revolution*. Cambridge MA:
    MIT Press, 2015.

Charmaz, Kathy. *Constructing Grounded Theory: A Practical Guide through Qualitative Analysis*. London: sage, 2006.

Clarke, Adele. *Situational Analysis: Grounded Theory after the Postmodern Turn*. London: sage, 2005.

Coombes, Brad, Jay T. Johnson, and Richard Howitt. "Indigenous Geographies III: Methodological Innovation and the Unsettling of Participatory Research." *Progress in Human Geography* 38, no. 6 (December 2014): 845–54.

Cope, Meghan. "Coding Qualitative Data." In *Qualitative Methodologies for Human Geographers*, edited by Ian Hay, 310–24. Oxford University Press, 2005.

Edwards, Paul N. *A Vast Machine: Computer Models, Climate Data, and the Politics of Global Warming*. Cambridge MA: MIT Press, 2010.

Eitzel, M. V., Emmanuel Mhike Hove, Jon Solera, Sikhangezile Madzoro, Abraham Changarara, Daniel Ndlovu, Adnomore Chirindira et al. "Sustainable Development as Successful Technology Transfer: Empowerment through Teaching, Learning, and Using Digital Participatory Mapping Techniques in Mazvihwa, Zimbabwe." *Development Engineering* 3 (2018): 196–208.

Eitzel, M. V., Jon Solera, Emmanuel Mhike Hove, K. B. Wilson, Abraham Mawere Ndlovu, Daniel Ndlovu, Abraham Changarara et al. "Assessing the Potential of Participatory Modeling for Decolonial Restoration of an Agro-Pastoral System in Rural Zimbabwe." *Citizen Science: Theory and Practice* 6, no. 1 (2021): 2. http://doi.org/10.5334/cstp.339.

Eitzel, M. V., Jon Solera, K. B. Wilson, Klever Neves, Aaron C. Fisher, André Veski, Oluwasola E. Omoju, Abraham Mawere Ndlovu, and Emmanuel Mhike Hove. "Indigenous Climate Adaptation Sovereignty in a Zimbabwean Agro-Pastoral System: Exploring Definitions of Sustainability 'Success' Using a Participatory Agent-Based Model." *Ecology & Society* 25, no. 4 (2020): 13.

———. "Using Mixed Methods to Construct and Analyze a Participatory Agent-Based Model of a Complex Zimbabwean Agro-Pastoral System." *PLOS ONE* 15, no. 8 (2020): e0237638.

Ellis, Carolyn, Tony E. Adams, and Arthur P. Bochner. "Autoethnography: An Overview." *Forum Qualitative Sozialforschung / Forum: Qualitative Social Research* 12, no. 1 (2011): 273–90.

Étienne, Michel. *Companion Modelling: A Participatory Approach to Support Sustainable Development*. Dordrecht, Netherlands: Springer, 2014.

Eubanks, Virginia. *Automating Inequality: How High-Tech Tools Profile, Police, and Punish the Poor*. New York: St. Martin's Press, 2018.

Freire, Paulo. *Pedagogy of the Oppressed*. New York: Bloomsbury Academic, 1970.

Goldman, Mara, Paul Nadasdy, and Matt Turner, eds. *Knowing Nature: Conversations at the Intersection of Political Ecology and Science Studies*. Chicago: University of Chicago Press, 2011.

Hampton, Stephanie E., Carly A. Strasser, Joshua J. Tewksbury, Wendy K. Gram, Amber E. Budden, Archer L. Batcheller, Clifford S. Duke, and John H. Porter. "Big Data and the Future of Ecology." *Frontiers in Ecology and the Environment* 11, no. 3 (April 2013): 156–62.

Haraway, Donna. "Situated Knowledges: The Science Question in Feminism and the Privilege of Partial Perspective." *Feminist Studies* 14, no. 3 (1988): 575–99.

Herr, Kathryn, and Gary L. Anderson. *The Action Research Dissertation: A Guide for Students and Faculty.* Thousand Oaks CA: SAGE, 2005.

Lave, Rebecca. "The Future of Environmental Expertise." *Annals of the Association of American Geographers* 105, no. 2 (March 2015): 244–52.

Merriam, Sharan B., and Elizabeth J. Tisdell. *Qualitative Research: A Guide to Design and Implementation.* 4th ed. San Francisco: Jossey-Bass, 2016.

Robbins, Paul. "Beyond Ground Truth: GIS and the Environmental Knowledge of Herders, Professional Foresters, and Other Traditional Communities." *Human Ecology* 31, no. 2 (2003): 233–53.

Salmond, Jennifer Ann, Marc Tadaki, and Mark Dickson. "Can Big Data Tame a 'Naughty' World?" *Canadian Geographer / Le Géographe Canadien* 61, no. 1 (March 2017): 52–63.

Star, Susan Leigh, and Karen Ruhleder. "Steps toward an Ecology of Infrastructure: Design and Access for Large Information Spaces." *Information Systems Research* 7, no. 1 (March 1996): 111–34.

Taylor, Peter J. *Unruly Complexity: Ecology, Interpretation, Engagement.* Chicago: University of Chicago Press, 2010.

Voinov, Alexey, Karen Jenni, Steven Gray, Nagesh Kolagani, Pierre D. Glynn, Pierre Bommel, Christina Prell et al. "Tools and Methods in Participatory Modeling: Selecting the Right Tool for the Job." *Environmental Modelling & Software* 109 (2018): 232–55.

Wilensky, Uri. *NetLogo.* Evanston IL: Center for Connected Learning and Computer-Based Modeling, Northwestern University, 1999. http://ccl.northwestern.edu/netlogo/.

Wilensky, Uri J., and Walter Stroup. "Learning through Participatory Simulations: Network-Based Design for Systems Learning in Classrooms." In *Proceedings of the 1999 Conference on Computer Support for Collaborative Learning,* edited by C. M. Hoadley and J. Roschelle, 667–76. Palo Alto CA: Stanford University, 1999.

Wilson, K. B. "Water Used to Be Scattered in the Landscape: Local Understandings of Soil Erosion and Land Use Planning in Southern Zimbabwe." *Environment & History,* no. 1 (1995): 281–96.

SEVEN
·······································

Citizen Scientists and Conservation in the Anthropocene

FROM MONITORING TO MAKING CORAL

Irus Braverman

Since the 1970s reef-building corals around the world have experienced a sharp decline as a result of the increasingly warming, acidifying, and polluted oceans. This decline became even more acute during the world's third-largest and longest bleaching event in 2014–17, when the Australian Great Barrier Reef, the largest living structure on Earth, lost more than half of its coral organisms, thus becoming the largest dying structure on Earth. Scientists project that most tropical corals around the world will experience annual bleaching events as early as thirty years from now and will not likely recover in human lifetime. The future of reef-building corals as we know them is looking very grim indeed, and coral reef scientists and laypersons alike are agonizing over how to save them. A fierce debate has emerged within the coral scientific community revolving around whether to focus efforts on traditional habitat preservation or to put more emphasis on restoration and what kind of restoration should be promoted.[1]

This chapter explores the involvement of citizen scientists in all three phases of coral conservation: monitoring and data collection, gardening, and reproductive restoration. First, in light of the growing occurrences of bleaching since the 1980s and the vast geographies of these occurrences, citizen scientists have been central to monitoring efforts, their work here mostly standardized and regulated by scientists. Second, for the last couple of decades nature enthusiasts, scuba divers, and local communities in various parts of the world have been involved in restoration efforts, with the underlying assumption that such efforts can make a difference for coral conservation. This is especially the case in the Caribbean, where coral decline was already registered in the 1970s, triggering multiple restoration efforts in response.

The last decade or so in particular has seen a flourishing of coral nurseries and restoration programs. This chapter describes the emergence of citizen science in coral reef restoration and argues that the growing field of reef restoration based on the coral-gardening method offers a unique opportunity for public engagement. By participating in these programs, citizen scientists have been moving beyond their traditional more passive role in data collection to actively restoring degraded species. At the same time, they have been doing so within the limited genetic confines of asexual reproduction: fragging (breaking) corals to make more corals of the same genetic kind and, typically, of a limited range of species.

Third, the chapter examines the impact of the growing field of home aquarists on how restoration is performed and the relationship between citizen scientists and experts. Science and technology studies researchers and theorists have generally interpreted the rise of citizen science as a sign of the democratization of science, since it enables people without professional credentials to produce scientific knowledge.[2] Researchers' focus on participation has sometimes led to criticism of citizen science projects that are not participatory enough—for example, if participants are not permitted to define the questions or analyze the data, or if their contributions are used for purposes not of their choosing.[3] Certain scholars have also called for "extreme citizen science," which emphasizes bottom-up, socially transformative projects with marginalized communities.[4] The chapter concludes by arguing that the increasing importance and changing role of citizen science in coral conservation—and participants' move toward a creative engagement that in fact challenges contemporary science—reflect a broader change in the role of the citizen scientists when it comes to nature conservation in the Anthropocene.

Traditional Data Collection in Coral Science

The goal of most citizen science projects is to answer specific scientific questions or to gather data to influence public policy.[5] In the context of corals, citizen scientists have played a major role in collecting and monitoring data that regional programs then use to assess coral reef status and trends, especially regarding coral bleaching (for instance,

Reef Check, REEF, BleachWatch, and CoralWatch).[6] Whereas scientific monitoring projects such as Coral Reef Watch (CRW), a program of the National Oceanic and Atmospheric Administration (NOAA), involve near-real-time and global measurement of sea surface temperature and are increasingly reliant on satellite data, there is also a growing need to perform "ground-truthing"—to humanly verify the data that has originated in a machine.[7]

This is a good example of the limits of digital infrastructures when it comes to something as fickle as coral health. But unlike the role that human sensors play in Smart Earth schemes that have been proliferating in other domains, coral restoration participants have arguably come to take a less prescribed and more intimate part in saving corals.[8] The documentary *Chasing Coral* illustrates this point when the cameras, which were planted in various parts of the ocean with much care and effort to document the corals' bleaching process, failed the data collectors. In their stead, the filmmakers had to send human divers with cameras to document the third global bleaching event in Australia's Great Barrier Reef.

On its website NOAA explains why it can do the satellite monitoring but is less capable of carrying out ground-truthing: "At this time, CRW does not have the resources to conduct in-water bleaching surveys. Instead, we rely on reports of coral reef conditions from partners and collaborators around the world to ground-truth our bleaching thermal stress measurements."[9] One of the many voluntary community-based partners that NOAA has relied on to generate data that verifies its own satellite data is CoralWatch.

CoralWatch has been documenting coral bleaching in more than sixty countries since 2002. Participants use a color swatch card to monitor coral bleaching and health (fig. 9), matching the chart colors to the coral color, recording the codes, and entering the data on a coral health chart (figs. 10 and 11). Scientists developed the colors on the chart using intentionally bleached coral in temperature-controlled aquariums.[10] As in many traditional citizen science projects, the scientists developed the standards and research strategies, and the laypersons are trained to follow their instructions and are relatively passive in the process. The active involvement of laypersons in restoration efforts

Welcome to CoralWatch

9. CoralWatch volunteer using a color-coded Coral Health Chart to document underwater bleaching. Courtesy of CoralWatch, University of Queensland, Australia.

can be thought of as a step toward a more democratic and collaborative engagement and knowledge production by citizen scientists, or what data scholars have also referred to as "citizen data."[11] Nonetheless, many coral scientists are questioning whether this citizen form of ecological engagement, novel as it may be, will really matter for the important project of saving tropical reef-building corals.

Coral Gardening

The recent realization of some coral scientists that preservation and monitoring corals might not save most of them has led to emphasis on more interventionist solutions to reef degradation, including coral-gardening practices. Coral gardening is used around the world by a growing number of restoration practitioners and local stakeholders.[12] Generally, it involves propagating coral stocks within coral nurseries and then "outplanting" them, which is the professional term used by the coral community for this practice, back onto degraded reefs.[13] Coral gardening has been expanding rapidly in Florida and the Caribbean, where hundreds of coral genotypes are propagated within in situ coral nurseries and thousands of coral colonies are outplanted onto degraded reefs each year.[14] Restoration sites range from locations selected for

How to Use the Coral Health Chart

1. Choose a random coral and select the lightest area.
2. Rotate the chart to find the closest colour match.
3. Record the colour code on a data slate.
4. Select the darkest area of the coral and record the matching colour code.
5. Record the coral type.
6. Continue your survey with other corals. Record at least 20 corals.
7. Submit your data using the CoralWatch Data Entry Apps or enter online at www.coralwatch.org.

Survey Methods include (depending on experience and location):
- **Random Survey** – Select corals randomly, such as the closest coral after every second fin kick.
- **Transect Survey** – Select corals by following a line (transect) and record every few meters.
- **Easily Identified Corals** – Select corals you can recognise and return to (permanent transect).

Tips
- Corals are fragile animals, make sure your survey has no affect on marine life.
- Due to colour loss at depth, use a torch when diving below 5 metres/15 feet.
- Avoid measuring growing tips of branching and plate corals since they are naturally white.
- Do not monitor blue or purple corals because they have a different bleaching response.
- Some corals are naturally lighter than others. Regular surveys are needed to look at coral health over time or pick up trends of bleaching and recovery.

Coral Types

| Boulder (BO) | Branching (BR) | Plate (PL) | Soft (SO) |

CoralWatch is a non-profit global coral reef monitoring program, working with volunteers worldwide to increase understanding of coral reefs, coral bleaching and climate change. **www.coralwatch.org**

10. "How to Use the Coral Health Chart" on the back of the CoralWatch chart used underwater by citizen scientists. Courtesy of CoralWatch, University of Queensland, Australia.

their ecological, cultural, or economic significance to miniprojects mushrooming at beach and island resorts around the world, where guests are invited to plant a coral or to sponsor a concrete coral frame with their name engraved on it.[15] Restoration projects vary not only in aims, geography, and target species but also in technique. Corals can be grown on metal cages, ropes, nets, and PVC trees; on the seabed; or in various positions in midwater (fig. 12).

The need to expand coral restoration programs, as well as the realization that education and outreach are central to the viability of these programs, has resulted in heightened engagement of local community members. A 2017 article by Dalton Hesley and colleagues focuses on the citizen science program Rescue a Reef.[16] Developed by the Univer-

Coral NO.	COLOUR CODE		CORAL TYPE (please tick)			
	Lightest	Darkest	Boulder	Branching	Plate	Soft
1						
2						
3						
4						
5						
6						
7						
8						
9						
10						
11						
12						
13						
14						
15						
16						
17						
18						
19						
20						

Your name:

Reef name:

Country:

GPS coordinates (if available):

Date:

Time:

Depth:
m / feet

Sea temp:
°C / °F

Sunny / cloudy / raining
Walking / snorkelling / diving
(please circle)

Your **DATA** is important to us, transfer your findings using 'CoralWatch Data Entry' Apps

▶ Google play App Store

or enter your data online at
WWW.CORALWATCH.ORG

Follow us

11. CoralWatch data slate information chart. Courtesy of CoralWatch, University of Queensland, Australia.

sity of Miami in 2015, the program aims to advance the conservation and restoration of Florida reefs by providing meaningful research, outreach, and educational opportunities to community members interested in contributing to the conservation of local marine resources.

The success of reef restoration and the recovery of threatened coral species requires both coral propagation and outplanting to be conducted at relevant ecological scales.[17] Restoration proponents have been arguing that the higher the number of outplanted corals, the higher the chance of natural recovery through sexual reproduction and the faster the recovery of lost ecological services provided by healthy and complex reef communities. Hesley and colleagues show that citizen scientists, after being provided a thirty-minute training lesson and an in-water demonstration, can make a significant contribution to reef

Braverman

12. Graduate student Elad Rachmilowitch adding new coral nubbins (minute coral fragments of one to several polyps) to the midwater bed nursery in Eilat, Israel, after the transplantation of grown corals. Each nubbin is glued to a plastic pin, and all the pins are inserted into quadrates of framed fine mesh. The nursery is situated six meters deep and twenty meters from the substrate. Photo by S. Shafir, 2012. Courtesy of Baruch Rinkevich.

restoration by outplanting threatened staghorn corals, which have the same survivorship rates as when they are placed on reefs by experts. Citizen scientists also typically cover their own expenses or contribute funds directly to the hosting programs, thus reducing the costs of restoration activities. This type of involvement provides hands-on education and outreach experience to the community.[18] For these reasons and others, some have argued that supplementing the field of reef restoration with citizen science offers a unique opportunity for public engagement in ecological recovery worldwide. By catalyzing local communities that are dependent on regional reef resources, the argument goes, citizen science projects can enhance the local economy.

Another example of local community involvement in restoration efforts is the Fiji-based nonprofit Corals for Conservation. The director of this program, Austin Bowden-Kerby, is renowned for his lead role in the documentary *The Coral Gardener*. Bowden-Kerby collects coral fragments from sites where they cannot grow for lack of space, transplants them onto raised platforms with healthy living conditions, and finally transplants them back onto the degraded reef. Over the years he has trained local fishermen to become "coral gardeners" like he is. These locals are hired by resorts to protect the reefs, which are declared as no-fishing zones. "The coral captures people; it captures their heart," Bowden-Kerby reflects in one of the film's scenes. "[But] once you have people's heart, what are you going to do with it? You're going to get them to work together to try and save the planet."[19]

Criticisms of Coral Gardening

The three major critiques of coral gardening focus on the cost, scale, and long-term relevance of the practice. In a 2016 review ecologist Elisa Bayraktarov and colleagues estimated the cost of restoring a hectare of reef habitat as exceeding US$150,000, higher than the cost of restoring a hectare of seagrass habitat and an order of magnitude higher than mangrove restoration.[20] Additionally, some have criticized the limited geographic scale of coral restoration. A recent fact sheet from a Florida nursery sets a goal of restoring one thousand acres, or one and a half square miles, of reef in the Florida Keys within the next ten years. However, the entire Florida Reef Tract is six hundred square miles, and the Great Barrier Reef—the world's largest—is more than two hundred times that size. Restoration efforts, critics argue, are simply inadequate, not only in scale but also in the identity of the corals being restored. In the United States, coral propagation and reef restoration programs have mostly focused on the threatened Caribbean coral genus *Acropora*. While important, this is only one of at least eighty-three coral species identified as imperiled by the International Union for Conservation of Nature.

A third criticism of coral restoration focuses on the underlying causes of degradation and, consequently, the long-term irrelevance of restoration. "I agree that restoration is better than doing nothing," environmental sociologist Joshua Cinner was recorded as saying to the press. "But your options aren't doing nothing or doing restoration. The millions of dollars spent on restoration could be spent improving water quality, managing fisheries, and reducing impacts of tourists. If [done well], these all can have tangible impacts on improving coral reefs."[21] As Stanford coral scientist Steve Palumbi puts it, "Regrowing corals won't work in areas in which the basic reason they died off in the first place has not been fixed. [There's] no point in trying to put a vegetable garden on a landfill."[22] If temperatures continue to climb, many scientists have stressed, no part of the ocean will be amenable to nurseries and replanting. Coral scientist Ove Hoegh-Guldberg is more explicit: "There's a whole industry that has been emerging around restoration. Unfortunately, I am the party pooper. Restoration won't work unless we stabilize the climate. I think we're wasting a lot of money doing this sort of thing. Not to say that we shouldn't be trying and refining the techniques and so on, but until we deal with the climate issue, [restoration] is futile. This is rearranging the chairs on the *Titanic* to get a better view."[23]

While they are more than willing to counter these specific critiques, advocates of coral restoration also point to the general agenda underlying them: scientists' apprehension about anything they deem unscientific. Restoration seems to trigger this apprehension both because of the high level of involvement by laypersons and practitioners, rather than scientists, and because the techniques used are mostly experimental and have not undergone rigorous scientific evaluation. Proponents of restoration are thus often not taken seriously by coral scientists, an observation that emerged clearly from my fieldwork at the International Coral Reef Symposium in Hawaii.[24] Amid this existing tension, coral aquarists recently entered the scene with an even more pronounced resistance against the exclusivity of the knowledge and datasets produced by coral scientists, mounting an outright challenge.

Coral Aquarists as Citizen Scientists

Coral husbandry has seen a steep rise since the 1990s with an increase in aquarium hobbyists. Contemporary hobbyists not only maintain corals for many years but also propagate and trade them.[25] Hobbyists usually take care of saltwater tanks—with live rock, anemones, corals, and other marine forms of life—in their private, typically urban homes.[26] Colin Foord of Coral Morphologic told me about the responsibility that comes with tending corals: "You're playing God [by] maintaining an entire trophic system. Unlike caring for a cat or a dog, you have to create and maintain this ecosystem on a 24/7 life support."[27]

For Foord this type of absolute care provides the best possible conservation education, especially for children. "The planet is really no different than an aquarium," he told me. Teaching children about the interdependencies of myriad factors within the tank's microecosystem also teaches them about caring for the planet. "How else can a landlocked kid acquire a sense of biophilia [love of life], especially toward organisms that don't have a cute face?" he asked rhetorically. In today's rapidly changing environments, humans have no other choice but to learn how to manage natural systems, Foord argued passionately.[28] Through practices of intensively caring for corals inside their urban homes and establishing a network of caregivers who share coral experiences among themselves, coral hobbyists have effectively become data citizens who engage one another through "data stories."[29]

Their engagement with corals within their own homes connects the individual hobbyists with a community of like-minded caregivers and with the ecological crisis of a dying ecosystem, shaping them into more involved participant political actors. Jennifer Gabrys, Helen Pritchard, and Benjamin Barratt observe in the context of air pollution sensing, "It is these practices that we see as constitutive of data citizenships, not as a designation of membership . . . , but rather as data-related engagements that activate political subjects and collectives in relation to environmental problems."[30] Rather than being quantified and standardized, the knowledge that the caretakers accumulate is dynamic, engaged, and relational; emerges out of their situated experiences; and is passed among them in the form of narratives. Thus these data citizen practices challenge not only the meaning of science but also the meaning of data.

The various groups of coral caretakers—whether biologists, divers, or hobbyists—share a love of and a bond with corals, as well as multiple story narratives, or data stories. Yet coral hobbyists are different from both coral biologists and scuba divers. While the former typically cannot travel away from their homes, as their corals require constant care, the latter travel to tropical sites around the globe that are mostly far removed from the city. The different groups also typically do not see eye to eye when it comes to more substantial issues. Historically, scientists and divers have blamed the coral hobbyists for much of the demise of wild reefs, pointing to the damage wrought on these systems by commercial exploitation for the aquarium trade.[31]

Prominent aquarist Julian Sprung told me in response that "scientists isolate themselves, often complaining that no one cares about their corals. But when they visit a coral aquarist show and see the level of enthusiasm toward corals, they are flabbergasted. They are used to thinking about our industry as a wasteland—a place where wild corals are taken to die—and are not aware of the huge mariculture tradition and the extent to which corals are actually alive and thriving in our homes."[32] "They used to point the finger toward us," Foord said. "Now people are realizing how damaging the carbon footprint of flying to these remote places is. It can't be overlooked—these divers might as well be burning coal."[33] The condescending approach of coral scientists, who have preferred to work in the field rather than inside a lab or in their own homes, has more recently turned into a recognition of the hobbyists' invaluable knowledge of coral husbandry and, furthermore, into early sparks of collaborative scientist-hobbyist efforts to save the dying reef-building corals of the world.

Since its emergence only two or three decades ago, the coral trade has become a multimillion-dollar industry.[34] A growing community of coral hobbyists has also emerged alongside it and is characterized by a tight social network on Facebook and a mushrooming of coral clubs around the country. According to Foord, over the past several years Facebook has become the most important platform for reef aquarists to ask questions, share photos of their tanks and corals, exchange information, and buy "frags" in auctions. "Many of the biggest online sellers use both Facebook and eBay to sell their coral frags," he told me. Web forums have been another important part of the coral hobby culture. Additionally, local reef clubs serve as central networks for the aquarium hobby. "Back in the 1990s there were very few saltwater-specific aquarium clubs, as most hobbyists used freshwater tanks back then," Foord recalled. Now, he added, even his rural home state of New Hampshire has its own reef aquarium club. Foord explained that "joining a reef club is probably the best way for a beginner or young person to acquire cheap coral frags or even colonies of fast growing, weedy 'basic' corals that aren't 'valuable' on the open market."[35] Large for-profit organizations such as Reef-A-Palooza and the Marine Aquarium Conference of North America orchestrate public coral events several times a year, attracting thousands of participants from all walks of life. A culture of caring for corals in the city, led by nonscientists far away from where conservationists care for them, has emerged.

Because much of the knowledge on captive coral husbandry was developed by aquarists and hobbyists rather than by biologists, the former have played a major role in the recent shift of restoration into the realm of coral husbandry in general and into that of sexual breeding in particular. Still, some aquarists and hobbyists have been frustrated by the lack of recognition of their contributions to the field by the coral scientists, who often consider their work as unscientific, rendering the data they gather and their knowledge invisible. "In my opinion," said Foord, "the real pioneers who cracked the code of coral growth and asexual reproduction in closed systems—the people who deserve the most credit for understanding the living biology of corals—are largely passionate amateur hobbyist aquarists who tinkered away in their basements and garages in the 1980s and worked out the chemical [and] physical needs of corals in a closed system. . . . Without them, there would still likely be no live coral exhibits at any public aquarium in the U.S. or Europe." He continued: "I would wager that there are more high school kids who are capably growing corals in their bedrooms than there are PhD coral biologists who would even know where to start. The difference in the relationship with coral between an old-school, gray-haired coral biologist and a coral hobbyist is that the coral [biologists] traditionally removed themselves . . . from the well-being of the coral. . . . But a hobbyist is emotionally, financially, and empathically connected to their coral. They live together twenty-four seven."[36]

Foord's narrative underscores that scientists are not the corals' exclusive caregivers. Moreover, from his perspective, hobbyists experience more intimacy with their subjects than do coral scientists. This narrative corresponds with that documented by Gabrys and colleagues: "By making, rather than merely accessing data, citizens generate distinct relations to types and uses of data, which can in turn be expressive of new data citizenship." Such novel relationships with data production and analysis are also referred to in the literature as "just good enough data" and "data justice."[37] Others similarly call for "a process for raising undisciplined lines of inquiry that are not constrained to the original research questions or approach."[38]

In their novel role as *conservation* aquarists, a growing number of former hobbyists are now performing central roles in restoration projects, such as building nurseries in imperiled marine areas and teaching biologists how to handle corals in labs. "Aquarists have discovered lots of stuff that biologists never saw before," Foord explained.[39] Founded and directed in 2002 by a former Rotterdam Zoo professional, the nonprofit SECORE (SExual COral REproduction) facilitates partnerships among aquariums and between their aquarists and marine biologists in the field. SECORE also executes large-scale restoration projects and develops strategies for increasing the corals' sexual reproduction for conservation purposes. Another important group is the Coral Restoration Foundation (CRF), which has emerged from the work of former live-rock grower Ken Nedimyer. The CRF is now managing the largest coral nurseries for restoration in the world and has been promoting the propagation of the *Acropora* hybrid, which is perceived as a controversial project by many coral scientists.[40] The layperson has thus transformed into the new expert, with newly recognized expertise that is grounded and engaged rather than abstract and standardized.

At the same time, one should note that both coral gardening and coral hobbyist practices are expressive of a particular *kind* of citizen science. As with another well-known version of citizen science, ornithology, it is mainly a white male European subject that encounters both a professionalizing science and a kind of hobby activity that must be made masculine by performing feats in the field, as Joeri Bruyninckx's work effectively shows.[41]

Conclusions

Traditionally, citizen science projects in the coral context have been less participatory and more contributory.[42] This chapter has traced the evolution of coral citizen science since the 1990s, from data collection and monitoring of bleaching to active participation and engagement and even into areas where nonscientists define the parameters of the field, informing the biologists about their discoveries. Drawing on the understandings advanced by nonexperts such as hobbyists and aquarists, local Indigenous groups have also been empowered to become stewards of their environments. But despite the intensifying involvement of the public in coral restoration, many scientists still question the feasibility and timeliness of this emerging field, suggesting that restoration is a mostly useless and even dangerously misleading practice.

Still, those who used to be relatively passive participants in coral conservation and whose participation was limited to data collection in accordance with scientific standards and regulations have become producers of knowledge, extending the realm of traditional data to consider alternative, more creative and engaged, ways of knowing.[43]

Notes

1. Braverman, *Coral Whisperers.*
2. Kimura and Kinchy, "Citizen Science."
3. Goodwin, "'Hired Hands'"; Ellis and Waterton, "Environmental Citizenship"; Lawrence, "'No Personal Motive?'"; Lakshminarayanan, "Using Citizens to Do Science"; Shirk et al., "Public Participation"; Haklay, "Citizen Science"; Irwin, "Citizen Science"; Cooper and Lewenstein, "Two Meanings."
4. Braverman, *Coral Whisperers.*
5. Hesley et al., "Citizen Science."
6. Branchini et al., "Using a Citizen Science Program"; Forrester et al., "Comparing Monitoring Data"; Marshall, Kleine, and Dean, "CoralWatch"; Roelfsema et al., "Citizen Science Approach."
7. CRW, "Daily Global 5km Satellite."
8. Bakker and Ritts, "Smart Earth."
9. CRW, "Monitoring Coral Bleaching."
10. Marshall, Kleine, and Dean, "CoralWatch."
11. Gabrys, Pritchard, and Barratt, "Just Good Enough Data."
12. Rinkevich, "Conservation of Coral Reefs"; Rinkevich, "Rebuilding Coral Reefs."
13. Rinkevich, "Restoration Strategies."
14. Schopmeyer et al., "Regional Restoration Benchmarks."
15. Braverman, *Coral Whisperers.*
16. Hesley et al., "Citizen Science."
17. Lirman and Schopmeyer, "Ecological Solutions."
18. Hesley et al., "Citizen Science."
19. Bowden-Kerby, *Coral Gardener.*

20. Bayraktarov et al., "Cost and Feasibility."

21. Rae, "For the Success of Coral Restoration."

22. Steve Palumbi (professor, Department of Biological Sciences, Stanford University, and director of Hopkins Marine Station), interview by author, June 24, 2016.

23. Ove Hoegh-Guldberg (director of Global Change Institute and professor of marine science, University of Queensland), Skype interview by author, February 25, 2015. Quoted in Braverman, *Coral Whisperers*, 58.

24. Braverman, *Coral Whisperers*.

25. Borneman, "Introduction to the Husbandry of Corals," 4.

26. Braverman, "Corals in the City."

27. Colin Foord (founder and codirector of Coral Morphologic), email message to author, November 18, 2017. Quoted in Braverman, "Corals in the City," 107.

28. Foord, Skype interview by author, December 27, 2017.

29. Gabrys, Pritchard, and Barratt, "Just Good Enough Data."

30. Gabrys, Pritchard, and Barratt, "Just Good Enough Data," 2; see also chapter 5.

31. Thornhill, *Ecological Impacts and Practices*.

32. Julian Sprung (aquarist), Skype interview by author, January 17, 2018. Quoted in Braverman, "Corals in the City," 104.

33. Foord, Skype interview by author, December 27, 2017.

34. Andrew Rhyne (assistant professor of marine biology), interview by author, May 11, 2016.

35. Foord, Skype interview by author, December 27, 2017.

36. Foord, Skype interview by author, December 27, 2017.

37. Gabrys, Pritchard, and Barratt, "Just Good Enough Data," 4; see also chapter 5; Taylor, "What Is Data Justice?," 6.

38. Moore et al., "Undisciplining Environmental Justice Research."

39. Foord, Skype interview by author, December 27, 2017.

40. Nicole Fogarty (assistant professor, Nova Southeastern University Oceanographic Center), Skype interview by author, June 15, 2017. See also Braverman, *Coral Whisperers*.

41. Bruyninckx, *Listening in the Field*.

42. Hesley et al., "Citizen Science," 95.

43. Gabrys, Pritchard, and Barratt, "Just Good Enough Data," 1.

Bibliography

Bakker, Karen, and Max Ritts. "Smart Earth: A Meta-Review and Implications for Environmental Governance." *Global Environmental Change* 52 (2018): 201–11.

Bayraktarov, Elisa, Megan I. Saunders, Sabah Abdullah, Morena Mills, Jutta Beher, Hugh P. Possingham, Peter J. Mumby, and Catherine E. Lovelock. "The Cost and Feasibility of Marine Coastal Restoration." *Ecological Applications* 26, no. 4 (2016): 1055–74.

Borneman, Eric. "Introduction to the Husbandry of Corals in Aquariums: A Review." In *Advances in Coral Husbandry in Public Aquariums*, edited by R. J. Leewis and Max Janse, 3–14. Arnhem, Netherlands: Burgers' Zoo, 2008. https://www.researchgate.net

/publication/228743590_Introduction_to_the_husbandry_of_corals_in_aquariums
_A_review.

Bowden-Kerby, Austin. *The Coral Gardener*. BBC, 2008, 9 min.

Branchini, Simone, Francesco Pensa, Patrizia Neri, Bianca Maria Tonucci, Lisa Mattielli, Anna Collavo, Maria Elena Sillingardi, Corrado Piccinetti, Francesco Zaccanti, and Stefano Goffredo. "Using a Citizen Science Program to Monitor Coral Reef Biodiversity through Space and Time." *Biodiversity and Conservation* 24, no. 2 (2015): 319–36.

Braverman, Irus. "Corals in the City: Cultivating Ocean Life in the Anthropocene." *Contemporary Social Science* (November 2019): 96–112.

———. *Coral Whisperers: Scientists on the Brink*. Oakland: University of California Press, 2018.

Bruyninckx, Joeri. *Listening in the Field: Recording and the Science of Birdsong*. Cambridge MA: MIT Press, 2018.

Cooper, Caren B., and Bruce V. Lewenstein. "Two Meanings of Citizen Science." In *The Rightful Place of Science: Citizen Science*, edited by Darlene Cavalier and Eric B. Kennedy, 51–62. Tempe AZ: Consortium for Science, Policy & Outcomes, 2016.

CRW (Coral Reef Watch). "Daily Global 5km Satellite Coral Bleaching Heat Stress Monitoring." 2018. https://coralreefwatch.noaa.gov/satellite/bleaching5km/index.php.

———. "Monitoring Coral Bleaching in the Field." 2018. https://coralreefwatch.noaa .gov/satellite/education/monitoring.php.

Ellis, Rebecca, and Claire Waterton. "Environmental Citizenship in the Making: The Participation of Volunteer Naturalists in UK Biological Recording and Biodiversity Policy." *Science and Public Policy* 31, no. 2 (2004): 95–105.

Forrester, Graham, Patricia Baily, Dennis Conetta, Linda Forrester, Elizabeth Kintzing, and Lianna Jarecki. "Comparing Monitoring Data Collected by Volunteers and Professionals Shows That Citizen Scientists Can Detect Long-Term Change on Coral Reefs." *Journal for Nature Conservation* 24 (April 2015): 1–9.

Gabrys, Jennifer, Helen Pritchard, and Benjamin Barratt. "Just Good Enough Data: Figuring Data Citizenships through Air Pollution Sensing and Data Stories." *Big Data & Society* 3, no. 2 (December 2016): 205395171667967.

Goodwin, Philip. "'Hired Hands' or 'Local Voice': Understandings and Experience of Local Participation in Conservation." *Transactions of the Institute of British Geographers* 23, no. 4 (1988): 481–99.

Haklay, Muki. "Citizen Science and Volunteered Geographic Information: Overview and Typology of Participation." In *Crowdsourcing Geographic Knowledge: Volunteered Geographic Information (VGI) in Theory and Practice*, edited by Daniel Sui, Sarah Elwood, and Michael Goodchild, 105–24. Dodrecht, Netherlands: Springer, 2013.

Hesley, D., D. Burdeno, C. Drury, S. Schopmeyer, and D. Lirman. "Citizen Science Benefits Coral Reef Restoration Activities." *Journal for Nature Conservation* 40 (2017): 94–99.

Irwin, Alan. "Citizen Science and Scientific Citizenship: Same Words, Different Meanings?" In *Science Communication Today: Current Strategies and Means of Action*, edited by Berhand Schiele, Joëlle Le Marec, and Patrick Baranger, 29–38. Nancy, France: Presses Universitaires de Nancy, 2015.

Kimura, Aya H., and Abby Kinchy. "Citizen Science: Probing the Virtues and Contexts of Participatory Research." *Engaging Science, Technology, and Society* 2 (2016): 331–61.

Lakshminarayanan, Shyamal. "Using Citizens to Do Science Versus Citizens as Scientists." *Ecology and Society* 12, no. 2 (2007): 2.

Lawrence, Anna. "'No Personal Motive?' Volunteers, Biodiversity, and the False Dichotomies of Participation." *Journal of Philosophy and Geography* 9, no. 3 (2006): 279–98.

Lirman, Diego, and Stephanie Schopmeyer. "Ecological Solutions to Reef Degradation: Optimizing Coral Reef Restoration in the Caribbean and Western Atlantic." *PeerJ* 4 (2016): e2597.

Marshall, N. Justin, Diana A. Kleine, and Angela J. Dean. "CoralWatch: Education, Monitoring, and Sustainability through Citizen Science." *Frontiers in Ecology and the Environment* 10, no. 6 (2012): 332–34.

Moore, Sarah A., Robert E. Roth, Heather Rosenfeld, Eric Nost, Kristen Vincent, Mohammed Rafi Arefin, and Tanya M. A. Buckingham. "Undisciplining Environmental Justice Research with Visual Storytelling." *Geoforum* 102 (2019): 267–77.

Rae, Haniya. "For the Success of Coral Restoration, a Matter of Scale." *Undark*, September 19, 2017. https://undark.org/article/coral-reefs-regrowth-restoration/.

Rinkevich, Baruch. "Conservation of Coral Reefs through Active Restoration Measures: Recent Approaches and Last Decade Progress." *Environmental Science and Technology* 39, no. 12 (2005): 4333–42.

———. "Rebuilding Coral Reefs: Does Active Reef Restoration Lead to Sustainable Reefs?" *Current Opinion in Environmental Sustainability* 7 (2014): 28–36.

———. "Restoration Strategies for Coral Reefs Damaged by Recreational Activities: The Use of Sexual and Asexual Recruits." *Restoration Ecology* 3, no. 4 (1995): 241–51.

Roelfsema, Chris, Ruth Thurstan, Maria Beger, Christine Dudgeon, Jennifer Loder, Eva Kovacs, Michele Gallo et al. "A Citizen Science Approach: A Detailed Ecological Assessment of Subtropical Reefs at Point Lookout, Australia." *PLOS ONE* 11, no. 10 (2016): e0163407.

Schopmeyer, Stephanie A., Diego Lirman, Erich Bartels, David S. Gilliam, Elizabeth A. Goergen, Sean P. Griffin, Meaghan E. Johnson, Caitlin Lustic, Kerry Maxwell, and Cory S. Walter. "Regional Restoration Benchmarks for *Acropora cervicornis*." *Coral Reefs* 36, no. 4 (2017): 1047–57.

Shirk, Jennifer L., Heidi L. Ballard, Candie C. Wilderman, Tina Phillips, Andrea Wiggins, Rebecca Jordan, Ellen McCallie et al. "Public Participation in Scientific Research: A Framework for Deliberate Design." *Ecology and Society* 17, no. 2 (2012): 29.

Taylor, Linnet. "What Is Data Justice? The Case for Connecting Digital Rights and Freedoms Globally." *Big Data & Society* (December 2017): 1–14.

Thornhill, Daniel J. *Ecological Impacts and Practices of the Coral Reef Wildlife Trade.* Washington DC: Defenders of Wildlife, 2012. https://defenders.org/sites/default/files/publications/ecological-impacts-and-practices-of-the-coral-reef-wildlife-trade.pdf.

EIGHT

Data Infrastructures, Indigenous Knowledge, and Environmental Observing in the Arctic

Noor Johnson, Colleen Strawhacker, and Peter Pulsifer

Environmental monitoring has become a key component of species and ecosystem management efforts and increasingly contributes to assessments of regional and global environmental change. In the Arctic region, where surface temperatures are increasing at twice the global average, observing and monitoring changes—for example, in species range distributions, glacial and sea ice extent and condition, and permafrost stability—now constitute a significant area of scientific coordination and investment.[1] Systems for monitoring include networks focusing on specific environmental zones and features, such as the International Network for Terrestrial Research and Monitoring in the Arctic, the Circumarctic Lakes Observation Network, and the World Glacier Monitoring Service. These diverse networks are united under the Sustaining Arctic Observing Networks (SAON), a "network of networks" established in 2016 to encourage "free, open, and timely access to high-quality data that will realize pan-Arctic and global value-added services and provide societal benefits."[2]

Much of the language associated with global and regional observing systems is highly technocratic; people are absent, save for passing references to "stakeholders" or the generically phrased "human dimensions" of observing. This modernist vision sees science as an inherent public good leading to societal benefits at national and global scales.[3] Society, in such a model, is divorced from science and from data collection; society is a beneficiary of data production but not involved in the process.

This vision is not unique to Arctic science. It can be seen in parallel forums such as the Group on Earth Observations (GEO), a collaboration of more than one hundred national governments and a similar number of participating organizations. GEO's mission is to build a Global Earth Observing System of Systems (GEOSS) as "a set of coordinated, independent Earth observation, information and processing systems that interact and provide access to diverse information for a broad range of users in both public and private sectors."[4] The graphic GEO uses to visualize GEOSS shows the planet surrounded by various deployed technologies (e.g., an airplane, satellite, balloon, and buoy) capable of capturing sensor-based data.[5]

The ways that people mediate the production of observations and data—and the values inherent in selecting, processing, interpreting, and applying the data—are rendered invisible in this vision of an earth-observing system. Also absent are details about specific data users, their information needs, and how data stored or shared through data infrastructures could be rendered into products that would meet user needs. To make sense of large datasets requires sophisticated algorithms and skilled mediation and interpretation on the part of data scientists. Their data work involves choices and interpretations that are inherently social and tend to reproduce power relations and inequalities already present within our systems of knowledge production and governance.[6]

Arctic observing networks reflect this social nature of data as they mobilize national science funding and channel scientific expertise and attention toward particular observing themes and topics. National funding agencies including the United States, Canada, the European Union, Russia, China, South Korea, and Japan have invested substantial resources in funding observational science and the development of Arctic observing networks. In 2016, at the first Arctic Science Ministerial, science ministers from eight Arctic nations considered the past, present, and future of science collaboration in the region. One of the priorities discussed was "the shared development of a science-driven, integrated Arctic-observing system that has mechanisms to maximize the potential of community-based observing and to draw on traditional and local knowledge."[7]

This ministerial priority reflects several assumptions that are shaping infrastructure development for observational data. First, it suggests that data holds intrinsic value, but that to realize this value, data must be managed through the development of large-scale systems. These systems draw on digital methods of storing, sharing, and archiving data and are designed to host diverse datasets provided by sensors and deployed technologies, creating what has variously been called Digital Earth, Program Earth, and Smart Earth, among other terms.[8] These infrastructures are envisioned as being able to accommodate and, in some cases, integrate highly diverse forms of data to make them usable by diverse stakeholders for the broad benefit of society.

A second assumption of the ministerial statement, perhaps somewhat more surprising given the emphasis in Smart Earth systems on remotely collected and digitally mediated data, is that Indigenous knowledge and community-based observing or monitoring (CBM) should be a central part of a large-scale observing system. This assumption reflects a growing interest in models of distributed data production that engage citizens as local observers and data collectors. Such contributions are sometimes referred to as "volunteered geographic information" when applied to digital infrastructures.[9]

In the Arctic residents conduct routine observations of the environment as part of situated practices that include hunting, fishing, camping, boating, snowmobiling, and collecting plants and berries. Programs developed to involve citizens in data collection have adopted a wide range of techniques, from training community members to collect data using scientific protocols or in situ sensors to more collaborative approaches in which citizens and visiting scientists codesign the monitoring protocols and cointerpret the data to support locally defined information needs.[10]

Some (though not all) of the observations collected by Arctic residents as part of CBM initiatives are based on Indigenous knowledge. When participating in formal observing and monitoring activities through organized programs, residents bring a deep knowledge of and experience in the local and regional environment and a personal sense of connection with that environment. The data products that result from these monitoring efforts range widely in content and form. Some are narratives, captured in sound or video files or transcribed interviews; others may be lists or notations about observed phenomena. Some projects have adopted technological interfaces such as handheld computers adapted for cold climates to facilitate easy collection of observations. Increasingly, CBM programs in the Arctic and elsewhere use cell phone apps for data collection. Many projects collect multiple forms of data, and some involve both locally contributed observations and remote-sensing data.

The emphasis in the Arctic region on the potential of Indigenous knowledge to contribute to broader observing systems is at least partly the result of Arctic Indigenous peoples' political movements. These movements have fought for land claims, for the right to self-determination through direct representation in the Arctic Council and other regional decision-making bodies, and as part of a global Indigenous peoples' movement advocating for recognition within the United Nations and other global bodies.[11] Two Inuit-led organizations, the Inuit Circumpolar Council and Inuit Tapiriit Kanatami, were the first to champion the inclusion of community-based monitoring as a contribution to SAON. CBM also reflects efforts to strengthen Indigenous sovereignty in national and subnational governance; decolonizing knowledge production and reclaiming research to ensure that it addresses needs and priorities of Arctic residents have been an important focus of these efforts.[12]

This chapter reviews the development of data infrastructures for Indigenous and community-based observations. We take inspiration from analyses of infrastructuring that emphasize the active and adaptive processes through which infrastructure is developed.[13] As Karen Baker reminds us, "This active form of infrastructure serves as a reminder that infrastructure is not just a thing but rather a set of arrangements, negotiations, and alignments that is a continuing state in terms of maintenance and update."[14] We draw on examples of collaborative infrastructure development from the Exchange for Local Observations and Knowledge of the Arctic (ELOKA), a program embedded within the National Snow and Ice Data Center (NSIDC) at the University of Colorado Boulder. ELOKA focuses on observations at a human scale, facilitating the collection, preservation, exchange, and use of Indigenous and local observations. In particular, we show how digital interfaces are adapted to address the specific needs of projects involving Indigenous knowledge. We consider this process of adapting data management infrastructures to meet the needs of Indigenous communities and Arctic residents to be one way of counteracting what Jim Thatcher and colleagues have termed "data colonialism," the processes of extracting and gaining profit from data by more powerful actors to the detriment of those with less power.[15]

Extending Arctic Data Infrastructure to Support Local Observations and Indigenous Knowledge

Scientists and data specialists at the NSIDC study the earth's dwindling snow and ice reserves, manage scientific data, and create tools and visualizations to allow researchers and the public to better understand the "world's frozen realms."[16] One of NSIDC's best-known data products is the "Arctic Sea Ice News and Analysis," which publishes daily and monthly maps and graphs of Arctic sea ice extent, along with a monthly narrative scientific analysis.[17] These images, based on remote-sensing data, compare current conditions with the median long-term average from 1981 to 2010. They are often used in scientific presentations about climate change because they offer a concrete visualization of dwindling Arctic sea ice.

Much of the data that NSIDC researchers manage is rendered through technological interfaces such as satellites and in situ sensors that measure geophysical phenomena like sea ice extent and thickness. These sensors collect data in numeric form that is turned into usable observations by researchers, who sort through and make meaning of numbers and translate the data into charts and graphs. This assemblage of sensors, computers, and scientists makes up a data infrastructure that delivers very specific data products that convey information about regional change in the Arctic. These visuals are shared via websites and scientific papers, reaching various users, who include academic and government scientists as well as science journalists. A *New York Times* article from March 2017, for example, drew on data from NSIDC to describe the new record low for the maximum Arctic sea ice extent.[18]

ELOKA fills a gap in Arctic science by providing support to community-based research and monitoring projects to ensure that the data and information they generate are protected and preserved, discoverable, and able to influence research, local decision-making, policy, and public awareness. ELOKA also seeks to build technical

Johnson, Strawhacker, Pulsifer

capacity within Indigenous peoples' organizations and to facilitate direct involvement with technical development, maintenance, and enhancement of data infrastructures. Finally, ELOKA works with community partners and collaborators to make sure that their data goals and aspirations are reflected in larger discussions about data within the broader Arctic research community. The next few sections briefly introduce three data infrastructure projects that ELOKA has supported through collaboration with community and Indigenous peoples' organizations and academic researchers. We then discuss the role of data infrastructure for local and Indigenous knowledge within the broader transformation of knowledge and expertise in Arctic environmental knowledge production.

Preserving and Sharing Place Names through the Yup'ik Environmental Knowledge Project

Calista Education and Culture Inc. (CECI) is a nonprofit organization representing the 1,900 Yup'ik tradition bearers of the Yukon-Kuskokwim delta in southwest Alaska. In 2000 CECI began a project to document Yup'ik knowledge, hosting a series of workshops with elders and youths from eleven communities. Out of these gatherings, documentation of place names became a theme of central importance to elders, who were worried about the implication of the loss of these names to younger generations. As participant Denis Shelden expressed, "If our young people especially forget about the land and the names and the hunting places and those rivers, it's like they will lose some of their body parts. But if they learn more about their identity, their minds will be stronger."[19]

CECI wanted to develop a website and an atlas as a way of sharing more than four thousand place names collected in these workshops to make them accessible to Yup'ik youths and ensure that they did not disappear with the passing of elders. These names are associated with a range of historic sites and geographic features, such as camp and settlement sites, rivers, sloughs, rocks, ponds, sandbars, and underwater channels.

Starting in 2010 ELOKA assisted with the development of a web-based atlas that geolocates the place names and incorporates embedded

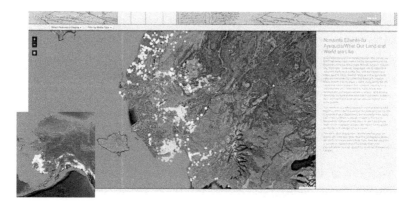

13. A screenshot of the Yup'ik Environmental Knowledge Project Atlas. Courtesy of authors (ELOKA).

sound files to assist with pronunciation of the Yup'ik language (see fig. 13). The atlas includes photographs, videos, and oral accounts for some of the sites. The site facilitates both long-term preservation and accessibility of data for younger Yup'ik, who live in rural villages and urban centers in Alaska and around the world. Although the atlas was initially password protected to address potential sensitivities associated with the data, representatives of the communities involved with the project voted unanimously to make the site and data open to the public.

The Yup'ik atlas infrastructure was developed to meet specific needs for data preservation and sharing that were identified through an active process of collaboration with community elders and youths. As Ann Fienup-Riordan, an anthropologist who works for CECI, explained, "Our work . . . did not begin as a mapping project. Rather, the places and their stories were what elders felt we needed to know. . . . Elders we traveled with immediately contextualized each site, noting the season during which people camped there, and how people traveled. Throughout they emphasized active ongoing relations between people and every aspect of their coastal environment."[20]

By choosing to make the site open-access, the elders allowed for the possibility that the information it contains might be useful to others and also created possibilities for new and unanticipated uses of the site as interests evolved over time. One potential use that was discussed was to document observations of environmental change within the landscape over time. More recently the collaboration expanded to include teachers and students from the Lower Kuskokwim School District in Alaska, who are adapting the atlas as a teaching resource to support delivery of a newly developed curriculum focused on Yup'ik values, language, and culture. Students are invited to contribute photos, video, audio, and text around three themes: families and communities, animals and plants, and the Yup'ik world and weather.

SIZONet and AAOKH, a Community-Based Observation Network

Another ELOKA partnership led to the development of the Local Observations digital database within the Seasonal Ice Zone Observing Network (SIZONet) and the Alaska Arctic Observatory and Knowledge Hub (AAOKH) (see fig. 14). These projects, led by researchers based at the University of Alaska Fairbanks, developed an "integrated program for observing seasonal ice in the context of a changing Arctic."[21] SIZONet and AAOKH facilitate the coproduction of sea ice knowledge by drawing on satellite data as well as observations of ice conditions and use contributed by Indigenous experts who serve as paid observers from nine villages on the northern and western coasts of Alaska. As researchers involved with the program note, Indigenous sea ice experts focus on different features of the sea ice environment than geophysicists do because they are engaged with this environment primarily through subsistence activities. Their observations focus on sea ice features, weather events, and animal behavior.[22] As diverse environmental and social changes are putting pressure on Indigenous knowledge systems, the goal of the observing program is to "archive, synthesize, and transmit observations of Indigenous experts during a time of rapid change."[23]

The program began by having Yup'ik and Inupiat hunters and sea ice experts keep logs of weather and ice conditions as they went about their activities on ice that is landfast, meaning it is anchored to the shoreline, or drifting. In addition to common variables such as wind speed and direction, the observers were encouraged to document details they felt were important related to hunting, community events, or environmental conditions and to use Indigenous terms. An adaptive database was developed that allowed researchers and community members to

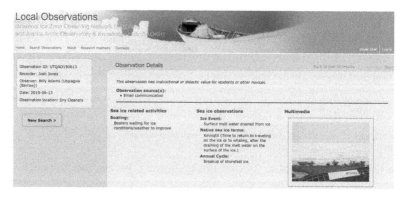

14. A screenshot of an observation record from the SIZONet/AAOKH catalog. Courtesy of authors (ELOKA).

identify important sea ice characteristics, processes, and events related to subsistence use and relevant to understanding geophysical changes. The database was updated to include new variables over time.

The SIZONet and AAOKH web interface is an example of data infrastructure that facilitates broader access to community-based observations. The interface allows the data to be interpreted more easily; by hosting data online and making it searchable in aggregate, community observers can track their own observations over time and compare them with those of other observers. The site allows keyword searches and displays results using icons for different features, such as ice, weather, boating, and specific animals like polar bear and seal.

Unlike the Yup'ik atlas, which focused on archiving and transmitting Yup'ik knowledge, this atlas aims to facilitate knowledge coproduction by making it possible to examine Indigenous observations of sea ice alongside geophysical data such as satellite images and sensor-based data. (The project maintains a separate website for sensor-based data that includes satellite images and data from ice observatories.) The project recognizes that knowledge contributed by community observers is contextual but also takes seriously the potential of this knowledge to contribute to broader understanding of sea ice and environmental change.

Because of this, the atlas is designed to facilitate ethical use of local observations by nonlocal guest users. To access the site, guest users

Johnson, Strawhacker, Pulsifer

15. Screenshot of the Atlas of Community-Based Monitoring and Indigenous Knowledge in a Changing Arctic (https://www.arcticcbm.org). Courtesy of authors.

must consent to a use agreement that asks them to acknowledge (by checking boxes) (1) that they will acknowledge and cite by name the person whose observations are being discussed; (2) that they will use a specific citation format provided; and (3) that interpretations of the observations will refer to the specific context in which the observations were made. A reference to an academic paper that describes this process is included to assist users. To protect potentially sensitive personal or cultural information, the complete observation logs are accessible only to community observers via secure access.[24]

A Circumpolar Inventory of Community-Based Monitoring Projects

The Atlas of Community-Based Monitoring and Indigenous Knowledge in a Changing Arctic is another initiative focused on data infrastructure for Arctic community-based observing (see fig. 15).[25] It began as a collaboration among the Inuit Circumpolar Council (ICC), Inuit Tapiriit Kanatami (ITK), and ELOKA. The first two organizations advocate for Inuit involvement in policy and decision-making within the Arctic Council and other international bodies (ICC) and within Canada (ITK). ITK's Inuit Qaujisarvingat, or Inuit Knowledge Centre, has a mandate to address issues related to decolonizing research practice and supporting Inuit community-led efforts for research and data management in Canada.

One of ICC's ongoing efforts is to ensure that Inuit knowledge is recognized as an important and valuable "way of knowing" and that it informs decision-making related to the Arctic. ICC's advocacy often describes Inuit knowledge as a separate but parallel body of knowledge that should be engaged along with conventional science. ICC was involved in the creation of SAON (described earlier in this chapter), serving on the Steering Committee and developing the atlas with the goal of ensuring that community-based monitoring and Indigenous knowledge would be part of this pan-Arctic network.

As an inventory of programs, the atlas was proposed as an initial step that would help determine the scale and scope of community-based monitoring programs in the region and support different kinds of network building. It would allow community-based observing programs that were monitoring similar phenomena in different parts of the Arctic to connect with one another in community-to-community networks, such as the one formed through the Siku Inuit Hila sea ice observation network.[26] It would also facilitate greater visibility of community-based programs within the broader network of researchers and government entities interested in Arctic observation and environmental change. Within the domain of pan-Arctic environmental initiatives, we hoped that the atlas would allow regional assessments—particularly those implemented by the working groups of the Arctic Council—to more easily solicit involvement by Indigenous knowledge holders and community-held data. We envisioned further development of the atlas to facilitate these kinds of data-sharing processes.

In developing the atlas, we conducted outreach to community-based monitoring programs identified through peer networks and academic and gray literature searches. One of the challenges we came up against was a lack of a shared terminology around what constitutes a community-based monitoring and observing program. A broad range of initiatives, from scientist-led programs aimed at collecting scientific data for research purposes to community-driven programs aimed at gathering information to support local decision-making, were added to the atlas. We decided to use a broad definition of CBM because part of the goal was to see the range of programs that existed; we saw development of a shared terminology or classification system as one potential next step based on this initial inventory process.[27]

Practitioners were asked to fill out a form with basic information about their programs, including what kind of phenomena they were monitoring, how monitoring was conducted, how data was stored and shared, whether it was used for decision-making purposes, and how community members were involved. The atlas was not intended to be a repository for community-based monitoring data, since many initiatives (like those described above) prefer to develop their own data-sharing infrastructures and are better equipped to adapt these infrastructures to protect sensitive information or to meet specific requirements of communities. The atlas could, however, be adapted to host data if programs expressed a need for and interest in this service.

Although the atlas does not host observational data, we treat the metadata gathered in the forms as data that requires review and consent from program leads before it can be added to the atlas. This limits the number of programs in the atlas and means that it is not a true inventory of CBM initiatives, as originally intended, but rather something between an inventory and the beginnings of an affiliative network. Whether programs chose to join the atlas was likely influenced by their degree of interest in connecting with knowledge and decision-making practices beyond the local scale. We discuss this in greater depth in the next section.

Reconfiguring Expertise through Data Infrastructure

Each of these three projects contributes in different ways to a broader movement toward knowledge and information sovereignty for Arctic Indigenous communities. Arctic residents express growing impatience with research practices that are disconnected from local goals and priorities. Many have strong concerns about researchers who have come, conducted interviews, and failed to inform them of the results of the research.[28] From their perspective, climate change has brought not only changes in the marine and terrestrial environment that have affected subsistence practices but also a growing interest on the part of researchers, who also affect the fragile ecosystem of Arctic communities in significant ways.[29]

These politics are mediated by formal institutions that have been developed at different scales of governance in the region. Calista Education and Culture Inc. is an example of such an institution operating at the community scale. Calista facilitates community-driven research projects and helps connect community interests in knowledge preservation with outside resources such as project funding and assistance with development of relevant data infrastructure. Projects like those of CECI are the easiest to evaluate as contributing to knowledge sovereignty, since they are defined and controlled at the local level, and community members have authority to determine how data is shared and who it is shared with. The data infrastructures developed at this level may be less likely to be compatible with larger ones; questions of interoperability are less prominent, since data is not intended to be used within broader analyses where other forms of data are also present.

Although implemented in Alaska, SIZONet and AAOKH are a good example of a regionally focused initiative that engages different kinds of knowledge, recognizing that community members and academic researchers have different perspectives on and bring different kinds of expertise to understanding the sea ice environment. In this case the data infrastructure that ELOKA assisted in developing was designed to facilitate analysis of local observations across a number of different locations and over time. This kind of comparison is much more difficult using traditional oral knowledge transmission methods. The online infrastructure is democratizing in that it facilitates access to and analysis of observations by everyone involved with the project, from community observers to academics, as well as opening up these resources to the public.

The infrastructure was designed around the specific needs of the project, including the need to recognize and maintain contextual elements of recorded observations, but it also has the potential for connection with similar efforts in other parts of the Arctic. ELOKA has worked to adapt the SIZONet and AAOKH interface for a community-based monitoring project in Greenland focused on natural resource use within the marine environment.[30] The political goal to which the project contributes could be described as one of local control over knowledge and data, but within a context of collaboration and copro-

Johnson, Strawhacker, Pulsifer

duction of knowledge. The data infrastructure requirements for this include the ability to build links between projects and to facilitate use of observations contributed by local observers within larger or broader scientific and decision-making processes.

The goal of facilitating the use of Indigenous knowledge and community-based observations within regional assessments and other processes animated the CBM atlas from its inception. ICC has emphasized the political importance of Inuit and other forms of Arctic Indigenous peoples' knowledge within the Arctic Council and its working groups. Thanks in large part to advocacy from Inuit leaders, the importance of traditional knowledge "to the collective understanding of the circumpolar Arctic" was officially recognized in the founding declaration of the Arctic Council.[31] As described earlier, the atlas was envisioned as a first step toward building an infrastructure that could facilitate better engagement of Indigenous knowledge and community observations at the regional level, based on the idea that visibility would facilitate network-building and collaboration. However, all partners recognize that a significant amount of work remains to operationalize the atlas as a platform for network building. We are currently seeking funding to support this next phase, which we imagine will involve developing mechanisms through which both data and expertise could be solicited to inform national, regional, or pan-Arctic decision-making and assessment processes.

Significantly, the infrastructure does not and cannot replace the need for interpersonal relationship building. Since it was not designed to host observations, the CBM atlas cannot facilitate direct access to community-based observational data. Rather, it can serve as a kind of broker, linking individuals interested in, for example, community-based observations of sea ice with programs that are gathering these kinds of observations. The value proposition that the atlas offers to community-based monitoring programs is external to the routine management of data that is now part and parcel of all kinds of research and monitoring efforts. Programs most interested in joining will therefore be those that already see themselves in relation to a broader Arctic observing network (which may not be true of many community-based programs, which tend to define themselves much more locally), as well as those whose community members have expressed a desire to share their knowledge with anyone interested in accessing it and those that would like to support the political project that is at the core of ICC's mission—namely, making a space for Indigenous knowledge within Arctic science and political decision-making.

Conclusions

ELOKA supports data infrastructure development for projects that are articulated within and across scales in the Arctic, from local to circumpolar. These projects deal with different kinds of data, from oral history and place name recordings to observation records with geospatial coordinates. They differ in their intended use, including whether they support sharing or contributing data beyond local or regional use. What they have in common is an interest in developing data infrastructures that can support ethical data management practices, help preserve local and Indigenous knowledge, and—when there is interest on both sides—connect these community initiatives to broader observing networks.

It is beyond the scope of this chapter to take up the question of how data infrastructures change the nature of the knowledge systems they engage. This is a question that has been the focus of considerable discussion in the academic literature on Indigenous or traditional knowledge, and for good reason. There is abundant literature documenting the ways that bureaucratic management and conventional scientific knowledge production co-opt, alter, or intentionally or inadvertently limit Indigenous knowledge systems, even in situations intended to facilitate inclusion or integration of such systems.[32] These structures are hard to break out of, and indeed, data infrastructures developed to transmit, archive, and share Indigenous knowledge necessarily simplify those knowledge systems.

For us, the question is not whether knowledge is transformed in the process of collection (as observations) and transmission (as data)—it is, of course. Rather, the questions that guide our engagement are how significantly communities are involved in shaping any particular project, and who is served by the data infrastructure. Knowledge systems in the Arctic are changing all the time, and data infrastructures need to be part

Johnson, Strawhacker, Pulsifer

of efforts aimed at preserving and transmitting Indigenous knowledge for the benefit of current and future generations. Such infrastructures can also facilitate collaboration among different knowledge systems. ELOKA's goal is to support cooperative efforts to advance understanding of the Arctic, based on ethical treatment of all knowledge.

Returning to the ministerial declaration calling for an Arctic observing system discussed at the outset of this chapter, we see several potential challenges that emerge from the vision it lays out. The stated goal is an "integrated" observing system that can "draw on traditional and local knowledge" and "maximize the potential of community-based observing." This vision of a singular, unified observing system is a lofty goal, but it comes across as proposing an instrumental use of community-based observing to achieve the goals of conventional science. Based on our experience, an integrated observing system is unlikely to succeed without first developing and strengthening infrastructures to support data management for Arctic communities. Integration is not the primary aim of most community observing efforts; inasmuch as some may like to see their observations valued and engaged in pan-Arctic networks, the primary motivation for observing is usually to meet local or regional information needs.

In spite of widespread interest in broad, integrative observing systems, such systems are still in their infancy. This means there is time to consider how to create flexible systems that can accommodate diversity. To be avoided is the adoption of a language and visualization of environmental observing on a global scale that erases the nuances of place—the environments in which people and animals make themselves at home. By erasing these specificities, the natural world is flattened into a range of phenomena that can be quantified and counted, and then observed and monitored from a distance. A better framework, and one that ELOKA is committed to supporting, is one that builds linkages and facilitates sharing of observations when feasible, but with the goal of developing many diverse observing networks. This should include multiple networks of local observers who may choose to orient their observations and develop data infrastructures to support them in very different ways. Such infrastructures can also facilitate collaboration between different knowledge systems by identifying complementarity between Indigenous and local knowledge and conventional science.

Acknowledgments

The authors would like to acknowledge the community members who have given their time and knowledge as partners in this work. We recognize the U.S. National Science Foundation, Office of Polar Programs, for supporting the ELOKA project (ARC 0856634, ARC 1231638, PLR 1513438).

Notes

1. Overland et al., "Surface Air Temperature."
2. SAON, "SAON Process."
3. Scott, *Seeing like a State*.
4. GEO, "About GEOSS."
5. Esri, "Face of Our Earth."
6. Thatcher, O'Sullivan, and Mahmoudi, "Data Colonialism through Accumulation."
7. White House, "Joint Statement of Ministers."
8. De Longueville et al., "Digital Earth's Nervous System"; Gabrys, *Program Earth*; Bakker and Ritts, "Smart Earth."
9. Sui, Elwood, and Goodchild, *Crowdsourcing Geographic Knowledge*.
10. Danielsen et al., "Environmental Monitoring."
11. Abele and Rodon, "Inuit Diplomacy"; Niezen, *Origins of Indigenism*.
12. Gearheard and Shirley, "Challenges."
13. Bowker et al., "Toward Information Infrastructure Studies."
14. Baker, *Informatics*, 11.
15. Thatcher, O'Sullivan, and Mahmoudi, "Data Colonialism through Accumulation."
16. NSIDC, "About Us."
17. NSIDC, "Arctic Sea Ice News and Analysis."
18. Fountain, "Arctic Winter Sea Ice."
19. Fienup-Riordan, "Linking Local and Global," 102.
20. Fienup-Riordan, "Linking Local and Global," 100.
21. ELOKA, "Seasonal Ice Zone Observing Network."
22. Eicken et al., "Framework and Database."
23. Eicken et al., "Framework and Database," 7.
24. Due to space constraints, we are not able to describe in detail the various mechanisms that ELOKA uses to adapt data infrastructures to the specific needs and sensitivities of Arctic communities. For a more thorough discussion of data management practices, see Pulsifer et al., "Role of Data Management."
25. Atlas of Community-Based Monitoring and Indigenous Knowledge in a Changing Arctic, https://www.arcticcbm.org.

Johnson, Strawhacker, Pulsifer

26. Huntington et al., "Community-Based Observation Programs."

27. Johnson et al., "Contributions."

28. Gearheard and Shirley, "Challenges."

29. Cameron, "Securing Indigenous Politics."

30. Danielsen et al., "Counting What Counts."

31. The terminology continues to be a point of discussion. ICC has proposed adoption of the term "Indigenous knowledge," while Arctic Council and its working groups have tended to use "traditional knowledge" or "traditional and local knowledge." Arctic Council, *Declaration*.

32. Agrawal, "Indigenous Knowledge"; Nadasdy, "Politics of TEK"; White, "Cultures in Collision."

Bibliography

Abele, Frances, and Thierry Rodon. "Inuit Diplomacy in the Global Era: The Strengths of Multilateral Internationalism." *Canadian Foreign Policy Journal* 13, no. 3 (2007): 45–63.

Agrawal, Arun. "Indigenous Knowledge and the Politics of Classification." *International Social Science Journal* 54, no. 173 (2002): 287–97.

Arctic Council. *Declaration on the Establishment of the Arctic Council*. Ottawa ON: Arctic Council, 1996. https://oaarchive.arctic-council.org/bitstream/handle/11374/85/EDOCS-1752-v2-ACMMCA00_Ottawa_1996_Founding_Declaration.PDF?sequence=5&isAllowed=y.

Baker, Karen S. *Informatics and the Environmental Sciences*. Scripps Institution of Oceanography Technical Report. San Diego: Scripps Institution of Oceanography, 2005. http://escholarship.org/uc/item/0179n650.

Bakker, Karen, and Max Ritts. "Smart Earth: A Meta-Review and Implications for Environmental Governance." *Global Environmental Change* 52 (2018): 201–11.

Bowker, Geoffrey C., Karen Baker, Frances Millerand, and David Ribes. "Toward Information Infrastructure Studies: Ways of Knowing in a Networked Environment." In *International Handbook of Internet Research*, edited by Jeremy Hunsinger, Lisbeth Klastrup, and Matthew M. Allen, 97–117. Dordrecht, Netherlands: Springer, 2010.

Cameron, Emilie S. "Securing Indigenous Politics: A Critique of the Vulnerability and Adaptation Approach to the Human Dimensions of Climate Change in the Canadian Arctic." *Global Environmental Change* 22, no. 1 (2012): 103–14.

Danielsen, Finn, Elmer Topp-Jørgensen, Nette Levermann, Piitaaraq Løvstrøm, Martin Schiøtz, Martin Enghoff, and Pâviârak Jakobsen. "Counting What Counts: Using Local Knowledge to Improve Arctic Resource Management." *Polar Geography* 37, no. 1 (2014): 69–91.

Danielsen, Finn, Neil D. Burgess, Per M. Jensen, and Karin Pirhofer-Walzl. "Environmental Monitoring: The Scale and Speed of Implementation Varies According to the Degree of Peoples Involvement." *Journal of Applied Ecology* 47, no. 6 (2010): 1166–68.

De Longueville, Bertrand, Alessandro Annoni, Sven Schade, Nicole Ostlaender, and Ceri Whitmore. "Digital Earth's Nervous System for Crisis Events: Real-Time Sensor

Web Enablement of Volunteered Geographic Information." *International Journal of Digital Earth* 3, no. 3 (2010): 242–59.

Eicken, Hajo, Mette Kaufman, Igor Krupnik, Peter Pulsifer, Leonard Apangalook, Paul Apangalook, Winton Weyapuk, and Joe Leavitt. "A Framework and Database for Community Sea Ice Observations in a Changing Arctic: An Alaskan Prototype for Multiple Users." *Polar Geography* 37, no. 1 (2014): 5–27.

ELOKA (Exchange for Local Observations and Knowledge of the Arctic). "The Seasonal Ice Zone Observing Network (SIZONet)." Accessed December 7, 2021. https://eloka -arctic.org/partner/seasonal-ice-zone-observation-network-sizonet.

Esri. "The Face of Our Earth: Esri's New Partnership with GEO/GEOSS." *Esri Insider*, December 10, 2015. https://www.esri.com/about/newsroom/insider/the-face-of-our -earth-esris-new-links-with-the-group-on-earth-observations-system-of-systems/.

Fienup-Riordan, Ann. "Linking Local and Global: Yup'ik Elders Working Together with One Mind." *Polar Geography* 37 (February 2014): 92–109.

Fountain, Henry. "Arctic Winter Sea Ice Drops to Its Lowest Recorded Level." *New York Times*, March 22, 2017. https://www.nytimes.com/2017/03/22/climate/arctic-winter -sea-ice-record-low-global-warming.html.

Gabrys, Jennifer. *Program Earth: Environmental Sensing Technology and the Making of a Computational Planet*. Minneapolis: University of Minnesota Press, 2016.

Gearheard, Shari, and Jamal Shirley. "Challenges in Community-Research Relation-ships: Learning from Natural Science in Nunavut." *Arctic* 60, no. 1 (2007): 62–74.

GEO (Group On Earth Observations). "About GEOSS." Accessed April 3, 2017. https:// www.earthobservations.org/geoss.php.

Huntington, Henry P., Shari Gearheard, Matthew Druckenmiller, and Andy Mahoney. "Community-Based Observation Programs and Indigenous and Local Sea Ice Knowl-edge." In *Sea Ice Field Research Techniques*, edited by Hajo Eicken, Rolf Gradinger, Maya Salganek, Kunio Shirasawa, Don Perovich, and Matti Leppäranta, 345–64. Fairbanks: University of Alaska Press, 2009.

Johnson, Noor, Lilian Alessa, Carolina Behe, Finn Danielsen, Shari Gearheard, Victoria Gofman-Wallingford, Andrew Kliskey et al. "The Contributions of Community-Based Monitoring and Traditional Knowledge to Arctic Observing Networks: Reflec-tions on the State of the Field." *Arctic* 68, no. 5 (2015): 1–13.

Nadasdy, Paul. "The Politics of TEK: Power and the 'Integration' of Knowledge." *Arctic Anthropology* 36, no. 1 (1999):1–18.

Niezen, Ronald. *The Origins of Indigenism: Human Rights and the Politics of Identity*. Berkeley: University of California Press, 2003.

NSIDC (National Snow and Ice Data Center). "About Us." Accessed April 3, 2017. https:// nsidc.org/about/overview.

———. "Arctic Sea Ice News and Analysis." Last updated December 2, 2021. http:// nsidc.org/arcticseaicenews/.

Overland, James E., Edward Hanna, Inger Hanssen-Bauer, Seong-Jung Kim, John E. Walsh, Muyin Wang, Uma S. Bhatt, and R. L. Thoman. "Surface Air Temperature." *Arctic Report Card: Update for 2018*. NOAA's Arctic Program. November 13, 2018. https://

arctic.noaa.gov/Report-Card/Report-Card-2018/ArtMID/7878/ArticleID/783
/Surface-Air-Temperature.

Pulsifer, Peter, Shari Gearheard, Henry P. Huntington, Mark A. Parsons, Christopher
McNeave, and Heidi S. McCann. "The Role of Data Management in Engaging Com-
munities in Arctic Research: Overview of the Exchange for Local Observations and
Knowledge of the Arctic (ELOKA)." *Polar Geography* 35, no. 3–4 (2012): 271–90.

SAON (Sustaining Arctic Observing Networks). "The SAON Process." Arctic Observing Network, accessed April 3, 2017. https://www.arcticobserving.org/background.

Scott, James C. *Seeing like a State: How Certain Schemes to Improve the Human Condition Have Failed.* New Haven CT: Yale University Press, 1998.

Sui, Daniel Z., Sarah Elwood, and Michael F. Goodchild, eds. *Crowdsourcing Geographic
Knowledge: Volunteered Geographic Information (VGI) in Theory and Practice.* Dordrecht, Netherlands: Springer, 2012.

Thatcher, Jim, David O'Sullivan, and Dillon Mahmoudi. "Data Colonialism through
Accumulation by Dispossession: New Metaphors for Daily Data." *Environment
and Planning D: Society and Space* 34, no. 6 (2016): 990–1006.

White, Graham. "Cultures in Collision: Traditional Knowledge and Euro-Canadian Governance Processes in Northern Land-Claims Boards." *Arctic* 59, no. 4 (2006): 401–14.

White House. "Joint Statement of Ministers on the Occasion of the First White House
Arctic Science Ministerial." September 9, 2016. https://obamawhitehouse.archives
.gov/the-press-office/2016/09/28/joint-statement-ministers.

NINE

Digital Infrastructure and the Affective Nature of Value in Belize

Patrick Gallagher

This chapter draws from a series of ethnographic encounters with geographic information system (GIS) scientists engaged in the "ground-truthing" of remotely sensed data. It focuses on Jonas, a GIS analyst from southern Spain working on a remote-sensing project in southern Belize.[1] Jonas is one component of a broad network of scientists, data technicians, and modelers who produce data-rich, spatially explicit digital representations of the Belizean landscape using GIS. The layers of environmental data that they collect and analyze provide the data infrastructure for models that are used to price the functions of nature as "ecosystem services." Advocates of this approach argue that mapping ecosystem services allows the material world to be appreciated as a form of "natural infrastructure" and that policymakers can use this understanding to make more informed decisions about conservation strategies and land use planning.[2]

This account attends to the mundane details of how Jonas collected and interpreted remotely sensed data for a Belizean landscape and highlights that this work of producing conservation's digital spatial infrastructures is imbued with both affect and ideology in a manner that fundamentally shapes how conservation knowledge, space, and value are made. Social scientists have noted that contemporary conservation governance is increasingly exercised through the use of these sophisticated spatial technologies, which often produce spaces of conservation interest by seeing the value of landscapes in utilitarian economic terms.[3] These technologies also represent a potent new interaction of traditional state technologies of governance (e.g., maps) with what have been termed neoliberal ways of exercising power (using the market to rationally regulate all aspects of social and political life). This chapter argues that the technical work of remotely sensing landscapes for GIS databases is in fact an affect-laden, embodied form of conservation labor. That is, conservation scientists make their remote technical work meaningful by engaging their bodies with the landscapes that they study and narrating the meaning of a particular landscape as they render it digitally. Their affective engagements with the landscape come to inform how they interpret and convey digital imagery. The chapter is situated at the intersection of studies concerned with performative models of nature and its economic value (and the labor of modeling), the social life of infrastructure, and the role of affect in shaping emergent forms of capitalist valuations of nature (sometimes referred to as ecosystem services).

Performative Models of Nature and Value

The global conservation movement has increasingly turned to market-oriented—or what critics refer to as neoliberal—methods in order to protect critical resources and habitats. This has involved a reconceptualization of the material environment as an economically valuable form of "natural capital" that renders services benefiting people.[4] This reorientation has led to the need for new ways of representing nature, its value, and how its value is distributed through space. Conservation scientists have frequently used GIS to create spatially explicit models of environmental value, including maps that can layer environmental data in a manner that brings into being new relational understandings of nature. For instance, a conservation GIS user might layer land use and land cover with various ecosystem functions such as water filtration or carbon sequestration to understand the spatial relationship between these variables.

Such models bring into being—literally perform—new kinds of nature and relationships between humans and nature. Drawing on earlier work on economic modeling, critical scholars of environmental modeling argue that these models, like models of the economy or climate systems, are not simply technologies of representation, but tools of governance that produce and manage conservation spaces and conservation markets in new ways.[5] Models are powerful symbolic tools that do not simply decipher a conservation landscape. Naomi Oreskes shows how "what we call data" may seem neutral, complete, and objective, but it is in fact partial.[6] Models of natural capital and ecosystem service value—so central to current conservation practice—come to "see" the forms of value in the landscape that their designers think are important and therefore include in the model. As a result, models may clearly represent a forest grove as providing economically valuable water filtration services but may be less likely to see or represent the grove as providing, for instance, cultural values such as spiritual sanctuary.

While much of this modeling occurs in a space that we might think of as "virtual," Donald MacKenzie argues that there is an important material dimension to the production of "virtual" models. "Science and technology," he says, "interact not as disembodied knowledge but as embodied expertise (often via the circulation of people)."[7] Environmental knowledge is embodied in both the material technologies of the conservation enterprise (like GIS) and the conservation scientists who labor to animate and make meaningful the representations produced by these technologies.[8] My ethnographic work with Jonas and other ecological modelers tasked with remotely sensing landscape data suggests a keen awareness of the limitations of abstract models and the embodied role of modelers in making these representations possible. The desire of modelers I worked with to engage in field-based ground-truthing of model data—and the narrative work they engaged in during these outings—are attempts to reimbue increasingly abstracted, virtual models with a sense of place and relational meaning that help animate their sometimes mundane infrastructural work. The labor of ground-truthing becomes an entry point into a space in between, in which a commitment to a powerful new model of nature's economic value can be maintained alongside a notion of nature as intrinsically valuable and therefore imbued with the emotions of attachment and care.

The Infrastructure of Market Nature

Infrastructure is defined in the social sciences as "built networks that facilitate the flow of goods, people, or ideas and allow for their exchange over space."[9] Infrastructure can refer to the traditional material structures of conveyance that we associate with the term, such as

Gallagher

roads, pipes, and wires. But it is increasingly also understood in digital terms, prompting social scientists to ask how emergent data technologies come to distribute valuable things and ideas and organize the relationships among people, organizations, and governments. Digital infrastructure is increasingly central to the work of environmental conservationists. Scientists concerned with the identification and conservation of natural capital and ecosystem services build a *digital* infrastructure to represent the material environment as a sort of *natural* infrastructure—as a space that when conserved and managed is capable of naturally doing the work of traditional built infrastructure (e.g., distributing water, purifying air and water, protecting from storm damage). The conservation scientists whom I worked with engaged with infrastructure in these two related forms. First, their database constituted a digital infrastructure for visualizing nature remotely. And second, this helped them represent the landscape as a site of naturally existing infrastructure.

Ethnographically engaging infrastructures, in their diverse forms, are one means of tracing the relationships among people and between people and the environment across space. In tracing these relationships, anthropologists have come to critically consider the underlying political and social visions that mobilize infrastructural projects.[10] Infrastructures can be understood as technical instantiations—material and digital—of an imagined ideal future, a mapping out of an ideal type of political vision for organizing worldly relations. Digital infrastructures for modeling the material environment, like all representations of nature, therefore convey a great deal about how their creators imagine social and political life to work.[11] In this view, rather than being a mere thing, infrastructure "can also . . . describe a sensibility: a way of thinking and acting in the world"—not just for its producers but also for the users who come to engage with it.[12] Infrastructure becomes an evocative symbolic object that functions both as a technology for distribution *and* as an object on which new distributive claims can be made.[13] The digital infrastructure for spatially valuing ecosystems makes possible new ways not just of representing nature's value but also of distributing it, by creating models that identify new forms of economic value in the material environment and thus newly valuable spaces.

The Affective Nature of Conservation Infrastructure

Infrastructural projects are therefore always more than technical: they reflect the desire to enact imagined futures based on idealized pasts, and the apparently technocratic is often made through affective registers of desire and nostalgia. Desire and nostalgia figure strongly in my engagement with conservation scientists and GIS professionals charged with developing rationalized models of nature as capital. Kathleen Stewart cites Raymond Williams's concept of the "structures of feeling" to argue that the affective captures the collected intimacies and intensities that shape everyday social life. Rather than a particular meaning or concrete experience, the affective captures the way in which these "social experiences in solution" come to "exert palpable pressures" in social life even as they escape clear "definition, classification or rationalizations."[14] These structures of feeling then shape and saturate even the most seemingly technical domains of social life.

Ordinary affect thus captures the ephemeral relationality of social life and the fleeting sensations by which relationships among people, and between people and their environment, come to take on meaning. These affective sensations and relations are embedded in infrastructure through the process of its often-mundane production, creating a technical and material project that often overflows with the feelings and desires of their producers. For conservation scientists engaged in spatial data collection and analysis, these affective overflows shape the experience of verifying remotely sensed environmental data during their embodied engagement with the material landscape. Ethnographic attention to these experiences of encounter reveals how technocratic models of nature as valuable infrastructure come to be made only through a series of affective moments in which the material environment provokes small but meaningful responses.

Ground-Truthing Pixels

I had met Jonas a few months earlier during a GIS workshop that he led for the World Wildlife Fund in far northern Belize. Jonas was a skilled GIS analyst who worked with a small but influential nonprofit conservation organization in the mountainous southern part of Belize.

The impressive data on land use and land cover that they produced was an important component in mapping and modeling ecosystem services in the region. Jonas invited me to join him in southern Belize to learn more about his remote-sensing work. On a clear, hot day Jonas and I took the bus north from his office in the town of Punta Gorda and north along the alluvial skirt of the Maya Mountains to a nondescript intersection with a dirt track road called the Blue River Turnoff. The road was used primarily by Mennonite farmers to reach their rural community on a broad savanna plain between the mountains and the sea. A small, squarely hewn, covered hitching post with a couple of bored, broad-framed horses marked the road's beginning.

Jonas and I were doing our fieldwork that day as part of a process generally termed ground-truthing. In GIS and cartography, ground-truthing refers to the process of verifying remotely sensed data (for example, from satellite imagery) by going to that site in physical space to observe what is actually there. Ground-truthing, Jonas told me, was necessary to make sure that what he was seeing was real. However, increasingly the process of ground-truthing does not actually involve the ground at all. Data accuracy is confirmed by comparing it with other sets of data. The "ground" comes to be produced through an algorithmically derived view from space. This lack of material engagement with the outside is frustrating to modelers like Jonas who were drawn to conservation work because of their passion for the outdoors.

Back in the office that afternoon Jonas showed me how he more typically experiences the landscape we had explored on foot earlier. On his office computer was a series of very high-definition Landsat satellite pictures of southern Belize. Because these satellites are equipped to detect light in eight different spectral bands, they are incredibly powerful tools for the remote sensing of land use and land cover. Infrared satellites are capable of detecting the levels of heat emitted from different land areas. The colors—ranging from black through the gray scale and into a subtle scale of pinks, climaxing finally in a penetrating deep pink—serve as a fairly effective proxy for the successional stages of a forest or field. These images are referred to as false color satellite images—false in that they do not represent the "true" perceived color of the landscape under examination, but instead are images that represent a broader spectrum of light than can be perceived by the human eye. These colors outside of the visible spectrum are represented in the images using an arbitrary set of color classes. The precise selection of colors used in false color composites can vary depending on what the user is trying to highlight, but for the mapping of land cover change, a combination of red and green is used.

The Landsat imaging system takes advantage of the distinctive spectral signature of vegetation. In false color images the near infrared band is typically assigned to be represented with red, so vegetation appears in various shades of pink and bright red. Quickly growing plants, usually new growth, reflect the greatest level of infrared light and thus appear a deep, bright pinkish red, while more mature, slower-growing vegetation appears in progressively lighter shades of pink. Areas of new growth—likely the result of the recent harvesting of older-growth trees—appear bright pink, standing out in the imagery as an intensely out-of-place threat to the landscape.

Jonas had been translating these tiny colored pixels, with the aid of a computer program, into a map indicating the stages of forest that could be found in this area of Belize. He had a remarkably intuitive sense for what he was viewing, with a capacity to put these endless pixels into a cohesive narrative of place and space. He would describe the relationships among various pixels through stories—stories about how people used the land, about hurricanes that tore down swaths of trees, about failed resort development projects and bad land deals, about growing Mennonite families and shifting Mayan cultivation practices.

Though the job of a GIS analyst is understood primarily in technical terms, his daily work came to be performed through a combination of his own notion of an inherent story to be told and the creative use of an elaborate technical apparatus to perform these narratives in image. He combined a sense for narrative with a strong technical capacity to then develop algorithms that essentially taught the computer program to share his intuition and translate the pixelated images into land cover maps automatically—that is, to spatially represent his narrative of land. This process of teaching the computer and formalizing his mode of image classification through code is referred to as training.

To train a computer to perform image classification, GIS users select a training site, a representative space in which they can carefully—and to their understanding, accurately—classify the different pixels as a model for the computer. The computer GIS program then associates different types of reflectivity with particular land use classifications and can replicate this work of classification automatically in other areas. This automated classification is known as supervised classification. The computer learns to autonomously do the work of land classification through careful training by its human supervisor.

The system classifies each pixel as a distinct and coherent classificatory unit. The pixel is the smallest graphic unit in the image. In Landsat images each pixel represents approximately thirty meters squared on the ground as a particular land use. Each pixel, based on its reflectivity, is translated into a land cover classification such as primary forest, savanna, or urban. Through this process Jonas and the computer GIS program produced a strikingly attractive and evocative color-coded map of the types of land use and land covers encountered in this area of Belize. These maps are important because by comparing the maps for a particular space through time, changes in land cover (and hence changes in measures such as carbon sequestration potential) over time can be measured quantitatively. Embedded within these quantitative measures, however, are a host of nuanced qualitative judgments that reflect particular understandings of nature, landscape, value, and change.

Following our day in the field, Jonas would export the georeferenced points that we collected as a layer of data for the GIS that he was developing. This layer of georeferenced data (or data linked to particular coordinates in space) would, if you imagine a transparency, be overlaid on the map of land use and land cover that he and the computer had been creating. The goal of our fieldwork then was to confirm that what we had encountered in the field correlated well with what Jonas, with the help of his computer, thought he was seeing on the satellite picture in the office. So, for instance, the first point that we collected on our walk was in an area that Jonas labeled at that time as "one year burn," meaning that it was a grassland area that had been burned within the past year. This area of fresh regrowth should have shown up on the infrared satellite as a bright pink and then have been

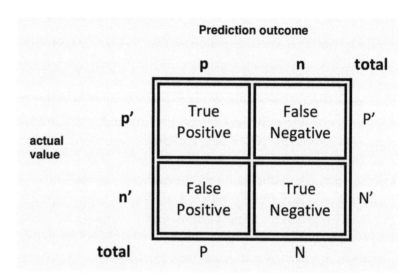

Prediction outcome

		p	n	total
actual value	**p′**	True Positive	False Negative	P′
	n′	False Positive	True Negative	N′
total		P	N	

16. A simple confusion matrix. A basic confusion matrix is set up like a statistical contingency table. GIS users can test the relative accuracy of their land cover maps by calculating the relationship between the predicted value (the classification of a pixel derived from initial remote sensing) and the actual value (the classification of a pixel derived from ground-truthing, or verification). Courtesy of author.

shaded on his land cover map accordingly (in this case, he chose yellow). If the first point did in fact fall within one of the areas colorcoded as new growth on the land cover map, then for this site at least, the accuracy of the satellite image and his system of pixel-by-pixel classification would be validated. By repeating this process numerous times with the help of a computer program, the accuracy of his work could be determined quantitatively.

The tool by which this ground-truthing is truthed is often referred to as a confusion matrix (see fig. 16). The confusion matrix "visually represents the difference between the actual and predicted classifications of a model."[15] The confusion matrix is a tool used in a variety of modeling applications, but the tool is also "frequently employed to organize and display information used to assess the thematic accuracy of a land-cover map."[16] It enables a quantifiable comparison of two sources of spatial information and is typically used in land use or land cover mapping to aid in the work of training the computer program to more accurately and automatically derive land classifications from maps. In

Gallagher

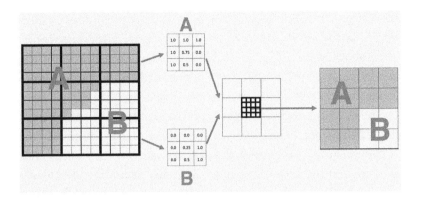

17. Diagram of the mixed pixel problem and the making of a crisp set. Each of the pixels (the large squares) is heterogeneous. This fuzziness can be resolved quantitatively by calculating the proportional makeup of each pixel. This simplifies analysis but obscures fine-grained landscape heterogeneity. Courtesy of author.

this case the matrix allowed Jonas to compare the guided classification of land cover that he performed with other scientists' classifications of the same or overlapping areas *or* to compare the computer-guided classifications with what we found in our fieldwork.

The most commonly used measure of overall accuracy for land cover map is the percentage of correctly allocated cases. But conventional accuracy assessments are based on what is termed the crisp set theory— which, according to a European Commission report on methods for land cover classification, is the idea that "the field to be mapped can be divided unambiguously into categories or themes" and "that each pixel in an image can be correctly allocated to a single theme" (fig. 17). The report notes dryly that "these simplifying assumptions help make much of the rigorous statistical analysis . . . possible."[17] Because of the limitations of assumptions built into crisp set theory—namely, that we can model as though the actual world is composed of a series of discrete pixels—there has been a push toward "fuzzy" or "soft" classification, which allows "for each pixel to have multiple and partial class membership."[18]

Using a standard crisp set analysis of correctly allocated pixels, Jonas confirmed that his classification of the tiny satellite pixels into a land cover type was correct approximately 86 percent of the time. The tech-

nical literature on ground-truthing suggests that this is a strong validation of Jonas's work (of course, the primary referent is other people's
attempts at classifications, so it's representations, or turtles, all the way
down). Yet it is the fuzziness of our unpixelated reality that concerned
Jonas and other GIS scientists most strongly.

Their skepticism of the binary approach of the crisp set analysis is
based on the fact that all wrong answers are treated as completely and
equally wrong. A crisp set analysis ignores the way each pixel is tied
into relations with those around it and how each pixel's meaning is
limited without understanding its context and its internal fuzziness.
A wrongly identified pixel might be understood somewhat more char-
itably with a fuzzy approach that considers the heterogeneity that frequently exists even at the relatively fine-grained scale of the pixel and
how that heterogeneity within one pixel links it to others.[19]

The pixels are, after all, arbitrarily brought into being by cameras
at random as they whip through space on satellites traveling at 4.735
miles per second. The world becomes pixelated by these orbiting cameras at random, shifting with each satellite orbit, and bears no relationship to how people socially and materially divide space on the
ground. But the GIS analysts and field ecologists who make use of these
remotely produced images for their analysis more commonly narrated
these landscapes to me as relationally produced social and ecological
spaces that they understand primarily through their embodied material and semiotic engagement. There is a tension, then, between their
desire to make meaning of conservation space through their fieldwork
and their desire to standardize the representation of that experience
through GIS technology. It is in their attempts to work through this
tension that much of the productive potential of GIS as a political and
social tool for its users lies.

In his ethnography of field ecologists in the Amazon, Hugh Raffles asserts that "an important politics is held in the gap between the
utopia of methodology and the awkward realities of field practice."[20]
Jonas and other field scientists in GIS who were keenly aware of the
frequent need for a fuzzier, more relational understanding of their
pixels were in some ways better positioned than most others to apply
a softer, fuzzy analysis to the broader project of GIS. Jonas enthusi-

astically consumed critical GIS literature that I would send along to him, literature that highlighted the limitations and assumptions about space and its representations built into the very structure of the software and its rules for making sense of and relating data. GIS, just one component of a broad suite of social technologies that come together to produce our images and understanding of nature, does not possess a single unambiguous quality or ethic by which its place in the project of making space, nature, and society can be classified. It is bound up in the infinite relations of contemporary conservation practice.

Scientists like Jonas and his colleagues used the ambiguity that was paradoxically produced through *crisp* set analysis to create new narrative spaces of nature and conservation. The binary representation of each pixel—either it has that attribute or it does not—is not typically the reality of that pixel. Within one pixel there can, in reality, be multiple land uses or land covers, and a combination of quantitative and qualitative judgments go into making a pixel crisp. GIS analysts then have a surprising degree of subjective influence in determining how pixels should be characterized. They frequently use their embodied understanding of the landscape to train models to identify pixels in a manner that aligns with their understanding of the landscape. That is to say, the constructed nature of crisp sets, in which each pixel must either possess or not possess a particular attribute, gives the analyst a degree of agency in how the transition from fuzzy to crisp is made. And with each iteration of the model, their affective ideas of landscape come to be embodied in the algorithms that animate these seemingly technocratic representations of nature and value.

Conclusions

The way that Jonas performed this fieldwork—stopping in at a small Mayan village, traversing a road made by Mennonite farmers, enlisting a local Maya forest reserve ranger to guide him, and inviting me to join the event so that he could perform his narration of those relations— was a means, intentional or subconscious, of engaging his technical infrastructural work with the network of relations and ideas that gave his work meaning. On our trip Jonas used the technical justifications for ground-truthing as an entry point into a broader practice of meaning making for the landscape—for his organization, for me, and perhaps most of all for himself. As Jonas talked about his work with me while walking the muddy track to Blue River, he revealed what may have been a more powerful impulse than simple data verification that drew him toward ground-truthing. That was the desire to go outside, to experience as real the space that through his labor had been reduced to a distanced digital abstraction. We went out, he said, "to make sure what I am seeing is real." When he said "what I am seeing," I think he was first referring to what he saw on the computer, but after spending the day with him, I also came to hear that as a broader mandate to see the world as he feels it really exists, to overcome what he felt was a fixation on abstraction in the infrastructural work of rendering nature valuable. Abstraction, in an effort to universalize notions of nature and its meaning, have actually made that meaning feel more tenuous for people like Jonas.

Ground-truthing was not simply about assuring a flat, one-to-one correspondence in the coding of a particular geographic coordinate by using two different techniques of classification. It was, for Jonas and other GIS technicians and analysts laboring to produce this new environmental infrastructure, a social and sensual process through which they aimed to take on a role that went beyond simply representing a landscape and toward engaging in its production. Jonas's field visit was merely one example of many cases in which conservation professionals described to me how GIS tools became ways they worked through affective—and not just technical—concerns with the meaning of socially valuable landscapes. Ground-truthing, in short, served two conflicting ends. It was a method for these conservation professionals to be *in* what they considered to be nature, to experience it and connect with it in ways that were fundamentally different from their everyday work with the GIS images. It was a desire that seemed to suggest a perceived shortcoming in the representative capacity of GIS. But by justifying these more sensual explorations of nature through a technical need to ground-truth, they were paradoxically producing streams of validating data that acted only to strengthen GIS's technical claim to be able to perform tremendous representational work for nature.

Jonas had learned, through expensive graduate training, to process nature through GIS because it was a form of environmental labor that was in tremendous demand. Skilled labor in GIS is desired because the technology is powerful for producing the abstracted representations of nature that have quickly become dominant in contemporary market-oriented conservation. It is the apotheosis of what Tim Ingold has called the "the global outlook," the notion that through technology we can look down on the globe as a whole environment, imagining and acting on the world from outside of it. Ingold argues that this "movement from spherical to global imagery is also one in which 'the world,' as we are taught it exists, is drawn ever further from the matrix of our lived experience," the notion that knowledge of our material world is best made through scientific tools that permit us to imagine leaving it.[21]

But to Jonas there were limits to the knowledge produced by leaving the places in which we dwell and make meaning. "There's no 'ultimate truth' in classification," Jonas said to me as we wrapped up the day's office work. You can only hope to improve the accuracy, and "you improve the accuracy with more and more people working in the area." But really, you need more than just more people working to map an area. "A lack of ground-truthing, *real on-the-ground* ground-truthing, is our biggest deficit," he continued. I could see it in his eyes right then: Jonas was already back in the field—as he did some final corrections to the classification colors in thirty-meter-square pixels one by one, his distant gaze suggested not boredom, as I had originally thought (or projected), but a maker of images imagining himself in each pixel.

Notes

1. A pseudonym has been used to preserve anonymity.
2. Aronson, Milton, and Blignaut, *Restoring Natural Capital*; Kareiva et al., *Natural Capital*.
3. Hazen and Harris, "Limits of Territorially-Focused Conservation"; Liverman and Cuesta, "Human Interactions."
4. Bakker, "Neoliberal Nature, Ecological Fixes"; Castree, "Neoliberalising Nature"; Daily et al., "Value of Nature"; Kareiva et al., *Natural Capital*.
5. Callon, "Some Elements of a Sociology"; Jasanoff, "Image and Imagination"; Lahsen, "Seductive Simulations?"; Oreskes, Shrader-Frechette, and Belitz, "Verification, Validation, and Confirmation."

6. Oreskes, Shrader-Frechette, and Belitz, "Verification, Validation, and Confirmation," 642.

7. MacKenzie, "Material Markets," 68. On the notion of virtualism, see Miller, "Turning Callon the Right Way Up."

8. Shapin, "Here and Everywhere."

9. Larkin, "Politics and Poetics," 328.

10. Anand, "Leaky States"; Appel and Kumar, "Finance"; Carse, "Nature as Infrastructure."

11. Worster, *Nature's Economy*.

12. Jackson et al., "Understanding Infrastructure."

13. Anand, *Hydraulic City*.

14. Stewart, *Ordinary Affects*, 3.

15. GIS Wiki, s.v. "Confusion Matrix," last modified September 26, 2016, http://wiki
.gis.com/wiki/index.php/Confusion_matrix.

16. Stehman, "Selecting and Interpreting Measures," 77.

17. Strahler et al., *Global Land Cover Validation*, 12.

18. Foody, "Status of Land Cover," 194.

19. Runk et al., "Landscapes, Legibility, and Conservation Planning." The authors further demonstrate that Landsat satellite imagery tends to obscure key human uses of the landscape, noting that the "selective harvest of large trees, most of the rice swiddens, and one homegarden site were not distinguishable from mature forest using Landsat satellite imagery and broad supervised classification methods" (172). Their findings emphasize that even "correct" classification fails to read important aspects of the social landscape and can contribute to impoverished understandings of the relationship between environment and society.

20. Raffles, *In Amazonia*, 169.

21. Ingold, "Globes and Spheres," 465.

Bibliography

Anand, Nikhil. *Hydraulic City: Water and the Infrastructures of Citizenship in Mumbai.* Durham NC: Duke University Press, 2017.

——. "Leaky States: Water Audits, Ignorance, and the Politics of Infrastructure." *Public Culture* 27, no. 2 (76) (May 2015): 305–30.

Appel, Hannah, and Mukul Kumar. "Finance." Theorizing the Contemporary, *Fieldsights*, September 24, 2015. https://culanth.org/fieldsights/finance.

Aronson, James, Suzanne J. Milton, and James N. Blignaut. *Restoring Natural Capital: Science, Business, and Practice.* Washington DC: Island Press, 2012.

Bakker, Karen. "Neoliberal Nature, Ecological Fixes, and the Pitfalls of Comparative Research." *Environment and Planning A* 41, no. 8 (2009): 1781–87.

Callon, Michel. "Some Elements of a Sociology of Translation: Domestication of the Scallops and the Fishermen of Saint Brieuc Bay." In *The Science Studies Reader*, edited by Mario Biagioli, 67–83. New York: Routledge, 1999.

Carse, Ashley. "Nature as Infrastructure: Making and Managing the Panama Canal Watershed." *Social Studies of Science* 42, no. 4 (2012): 539–63.

Castree, Noel. "Neoliberalising Nature: The Logics of Deregulation and Reregulation." *Environment and Planning A* 40, no. 1 (2008): 131–52.

Daily, Gretchen C., Tore Söderqvist, Sara Aniyar, Kenneth Arrow, Partha Dasgupta, Paul R. Ehrlich, Carl Folke et al. "The Value of Nature and the Nature of Value." *Science* 289, no. 5478 (July 2000): 395–96.

Foody, Giles M. "Status of Land Cover Classification Accuracy Assessment." *Remote Sensing of Environment* 80, no. 1 (April 2002): 185–201.

Hazen, Helen D., and Leila M. Harris. "Limits of Territorially-Focused Conservation: A Critical Assessment Based on Cartographic and Geographic Approaches." *Environmental Conservation* 34, no. 4 (December 2007): 280–90.

Ingold, Tim. "Globes and Spheres: The Topology of Environmentalism." In *Environmental Anthropology: A Historical Reader*, edited by Michael Dove and Carol Carpenter, 462–70. Malden MA: Blackwell, 2007.

Jackson, Steven J., Paul N. Edwards, Geoffrey C. Bowker, and Cory P. Knobel. "Understanding Infrastructure: History, Heuristics and Cyberinfrastructure Policy." *First Monday* 12, no. 6 (June 2007). http://www.firstmonday.dk/ojs/index.php/fm/article/view/1904.

Jasanoff, Sheila. "Image and Imagination: The Formation of Global Environmental Consciousness." In *Changing the Atmosphere: Expert Knowledge and Environmental Governance*. Cambridge MA: MIT Press, 2001.

Kareiva, Peter, Heather Tallis, Taylor H. Ricketts, Gretchen C. Daily, and Stephen Polasky. *Natural Capital: Theory and Practice of Mapping Ecosystem Services*. New York: Oxford University Press, 2011.

Lahsen, Myanna. "Seductive Simulations? Uncertainty Distribution Around Climate Models." *Social Studies of Science* 35, no. 6 (December 2005): 895–922.

Larkin, Brian. "The Politics and Poetics of Infrastructure." *Annual Review of Anthropology* 42, no. 1 (2013): 327–43.

Liverman, Diana M., and Rosa Maria Roman Cuesta. "Human Interactions with the Earth System: People and Pixels Revisited." *Earth Surface Processes and Landforms* 33, no. 9 (2008): 1458–71.

MacKenzie, Donald. *Material Markets: How Economic Agents Are Constructed*. Oxford: Oxford University Press, 2009.

Miller, Daniel. "Turning Callon the Right Way Up." *Economy and Society* 31, no. 2 (January 2002): 218–33.

Oreskes, Naomi, Kristin Shrader-Frechette, and Kenneth Belitz. "Verification, Validation, and Confirmation of Numerical Models in the Earth Sciences." *Science* 263, no. 5147 (February 1994): 641–46.

Raffles, Hugh. *In Amazonia: A Natural History*. Princeton NJ: Princeton University Press, 2002.

Runk, J. Velásquez, Gervacio Ortíz Negría, Leonardo Peña Conquista, Gelo Mejía Peña, Frecier Peña Cheucarama, and Yani Cheucarama Chiripua. "Landscapes, Legibility, and Conservation Planning: Multiple Representations of Forest Use in Panama." *Conservation Letters* 3 (2010): 167–76.

Shapin, Steven. "Here and Everywhere: Sociology of Scientific Knowledge." *Annual Review of Sociology* 21, no. 1 (August 1995): 289–321.

Stehman, Stephen V. "Selecting and Interpreting Measures of Thematic Classification Accuracy." *Remote Sensing of Environment* 62, no. 1 (October 1997): 77–89.

Stewart, Kathleen. *Ordinary Affects*. Durham NC: Duke University Press Books, 2007.

Strahler, Alan H., Luigi Boschetti, Giles M. Foody, Mark A. Friedl, Matthew C. Hansen, Martin Herold, Philippe Mayaux, Jeffrey T. Morisette, Stephen V. Stehman, and Curtis E. Woodcock. *Global Land Cover Validation: Recommendations for Evaluation and Accuracy Assessment of Global Land Cover Maps*. Luxembourg: Office for Official Publications of the European Communities, 2006. https://op.europa.eu /en/publication-detail/-/publication/52730469-6bc9-47a9-b486-5e2662629976 /language-en/format-PDF/source-117104416.

Worster, Donald. *Nature's Economy: The Roots of Ecology*. San Francisco: Sierra Club Books, 1977.

TEN

Infrastructuring Environmental Data Justice

Dawn Walker, Eric Nost, Aaron Lemelin, Rebecca Lave, Lindsey Dillon,
and Environmental Data and Governance Initiative (EDGI)

The United States' authoritarian tendencies demonstrated by and reproduced in the wake of the 2016 federal elections manifested in environmental policy and the data infrastructures that surround it. A series of executive orders by President Trump and directives from U.S. Environmental Protection Agency (EPA) administrators desperately sought to shore up the structures of racial capitalism by dispossessing marginalized communities of their environmental and informational rights. Such efforts reflected intensified corporate influence on regulatory processes and included introducing lobbyists to—and removing staff scientists from—EPA advisory boards; the creation of a panel to question established climate data, a proposal that would have effectively prevented the EPA from using public health data in rule-making; and cuts to environmental programs that protect minorities and vulnerable populations. While President Biden's administration subsequently responded to these actions with a larger budget for the EPA, centering environmental and climate justice in policy, and declaring its faith in science, such responses do not necessarily undo long-standing declines in funding, capacity, and accountability that the Trump administration exacerbated, raising questions about the value and integrity of federal environmental information and concerns about public access to data and the continuity of data collection.

In response to these political challenges, a consensus-based, geographically distributed organization of academics, professionals, and organizers known as the Environmental Data and Governance Initiative (EDGI) formed in November 2016. Since then EDGI has worked to document, contextualize, and analyze environmental data and governance practices at the federal level in the United States. EDGI's projects include archiving datasets, convening online workshops around trends in enforcement and compliance, interviewing EPA and Occupational Safety and Health Administration employees about conditions inside those agencies, and monitoring federal environmental agency websites for changes in access and content. EDGI has written a variety of reports on the Trump administration's strategies for undermining the EPA, its record of removing and altering important web-based environmental information and resources, and current threats to environmental justice (EJ) and opportunities to reimagine it in Green New Deal proposals.[1]

This chapter reflects on the changing nature of EDGI's collaborative data-archiving work to shed light on the political nature of data infrastructures. EDGI's data archiving has shifted from a project aiming to save government environmental data to a broader project of rethinking the infrastructures required for community stewardship of data. We describe the DataRescue movement, which began in December 2016 at a Guerilla Archiving event hosted by the Technoscience Research Unit at the University of Toronto. As crucial as DataRescue has been for raising awareness about the vulnerability of government data, it also raised questions for us about who benefits from data and its stewardship. The focus on *saving* existing data meant that DataRescue events did not address broader questions of how and why this particular data is collected by federal agencies in the first place.[2] Taking these concerns seriously led us to rethink our approach going forward and ask: How can we build the social and technical infrastructures to make data and decision-making more accessible, accountable, and, ultimately, just?

We draw inspiration from EJ and data justice (DJ) scholars and organizers concerned with the politics of representation and knowledge to extend our preliminary thinking around "environmental data justice" (EDJ).[3] These areas of scholarship demonstrate how data is embedded in historical, political, economic, and social systems of oppression, and they outline tactics to enact alternatives. Informed by these moves, we make the case for connecting the emerging framework of data justice to environmental justice in order to envision more equitable futures. We conclude by reflecting on Data Together, a discussion space EDGI facilitates with partners from the tech industry. In Data Together we envision new infrastructures for environmental data, including those enabling decentralized web page and dataset archiving. Using the notion of ED J, we reflect on active tensions in Data Together.

Collaborative Data Rescue

In the aftermath of the 2016 U.S. presidential election, there was widespread public concern that the Trump administration would seek to eliminate or alter web pages and datasets, given the actions of the Bush administration to hinder access to chemical safety data after 9/11 and the Harper government's (2006–15) actions against scientists in Canada.[4] Mobilizing quickly, local organizers hosted forty-nine DataRescue events to archive U.S. federal environmental web pages and datasets in cities across the United States and Canada between December 2016 and June 2017, with support from ED GI and the DataRefuge project at the University of Pennsylvania. The first of these public Guerilla Archiving events was held on December 17, 2016, at the University of Toronto, organized by the Technoscience Research Unit and ED GI (fig. 18). At that event more than 150 scholars, students, technologists, and activists gathered to nominate key datasets for inclusion in the Internet Archive's Wayback Machine. The Internet Archive is a nonprofit digital library and creator of the Wayback Machine, a browser-based application that can play back snapshots from different moments of the archive's more than 308 billion preserved web pages. The University of Toronto event also sought to strategize how to deal with links and datasets that would not be preserved through available automated methods.

Over the course of these events, DataRescue attendees nominated over sixty-three thousand web pages as "seeds" using a custom browser extension developed before the first event. In addition, more than twenty-two thousand datasets were identified as candidates for non-automated preservation. Several hundred of these were harvested through a workflow developed by ED GI and the DataRefuge project and uploaded to the DataRefuge repository. One criterion in identifying which datasets and web pages to archive was their importance to EJ activism, particularly datasets that provided a user-friendly interface to aid understanding of the prevalence of toxicants in commu-

18. Participants at the first Data Rescue event, in December 2016 at the University of Toronto. Courtesy of authors.

nities. For example, early on we archived the E P A's Enforcement and Compliance History Online portal, which collates data on industry emissions and regulatory actions.

When DataRescue participants nominated a page as a seed, they were instructing the Internet Archive to save a copy of that specific page and also to start crawling, or systematically visiting web pages and following links, making copies of intermediate pages along the way. However, crawler software is not able to fully archive and discover links to datasets and web pages on all sites, partially because of the underlying web technologies and internet infrastructure and partially because of resource and storage constraints; as a result, not everything is discovered or meaningfully archived. This means web pages including the E P A Pesticide Chemical Search, which provides information and access to data on chemicals only after queries are submitted by chemical name or registration number, would not be meaningfully archived. At many events attendees developed custom solutions in cases like this to scrape, or extract, links and datasets that the crawler likely would

Walker et al.

not automatically process. To aid in this process EDGI members created a tool kit to support local organizers, a categorization system for federal agency sites to coordinate across events, and scripts for custom scraping solutions. The evolving workflow developed in the DataRefuge project addressed chain of custody and metadata considerations, seeking to provide verification, as well as citational ability for these community-preserved datasets if the originals were no longer accessible.

In many ways EDGI's early archiving work represented a response to a crisis as we rapidly prepared for worst-case scenarios. Thousands of people connected at DataRescue events over a shared concern about threats to environmental and climate data. By trying to *save* existing data, however, DataRescue events ultimately did not address broader questions around the logic of data extraction that is at the heart of why federal agencies collected this data in the first place and how they use it.[5] Further, EDGI's archiving did not account for either the scale of the data or the ongoing and complex kinds of technical and policy considerations that their digital preservation and stewardship involved. Because existing federal records laws were drafted before the invention of the internet, but most government information today is "born digital," what DataRescue attempted was socially and technically impossible.[6]

We regrouped at an online DataRescue Town Hall on April 1, 2017, where we celebrated our shared accomplishments, identified challenges, and heard feedback from event organizers. Within EDGI we reflected collectively about our role in facilitating rescue events as well as possibilities for any broader movement going forward. By setting our sights on all U.S. environmental and climate data across all federal agencies, we had defined a project large enough to get lost in. Approaching a large-scale and distributed project meant there was no easy way to determine when we would be done. We could not look at a single catalog and identify all the federally published data we needed to "rescue," because even a seemingly comprehensive catalog like data.gov is incomplete. Finally, we recognized our archiving workflow required a large amount of continued participation to balance collaboration between those with interest or enthusiasm and those with domain expertise.

As an organization, EDGI begins from the position that data is not inherently good, or even neutral, and yet our initial work in DataRes-cue did not necessarily reflect this commitment. During the coordi-nated rescuing of data, we found it difficult to continue our previous critiques of the politics of data collection and stewardship in which we would ask how that data had originally been collected, how it had been used to advance agencies' interests, or how it might be made use-ful by different communities. Taking these concerns seriously led us to rethink our work and to develop a set of practices we have termed environmental data justice (EDJ), which center on providing just access to, interpretation of, and control of data as important goals in and of themselves and as means to broader socioenvironmental transforma-tions. As EDGI has moved from DataRescue to Data Together and specific projects inspired by it, we have sought to bring scholarship on environmental justice and data justice together in productive con-versation through the framework and practices of EDJ.

Critical Interventions on Data

EJ scholars and organizers have grappled with the politics of environ-mental data for decades. They have challenged exclusions from environ-mental knowledge production, asking who gets to collect authoritative environmental data and how. They have also produced alternative forms of environmental data, since most information about pollution in the United States is self-reported by industry or reliant on inadequate or incomplete collection methods.[7]

In one example from the 1990s, organizers from a neighborhood on the border of a Shell Chemical facility in Louisiana developed a low-cost method of air quality monitoring, called "air buckets." They did so because they discovered that the state—which had declared Shell Chemical's air quality emissions safe—was relying on inadequate data about the air they breathed.[8] The state was collecting data on the aver-age of toxic chemicals over twenty-four-hour periods and comparing these with ambient air standards. Louisiana organizers, who called themselves the Louisiana Bucket Brigade, argued that these averages flattened out short-term spikes in air pollution levels and therefore obscured the moments they were most at risk. Instead, the Bucket Bri-gade used bucket sampling to rapidly collect data on air quality during

Walker et al.

these moments of peak emissions. In doing so they demonstrated the capacity of fenceline communities adjacent to polluters to produce scientific knowledge about air quality and improve environmental decision-making. Through these and other campaigns, EJ advocates have questioned, "What counts as data, what data are collected, and whose interests do they serve?"[9]

Likewise, DJ scholars have called for rethinking data science within a social justice framework that asks: Whom does data benefit? Research has emphasized the disciplining aspects of data, including how data surveillance constrains social movements, how data-driven governance entrenches power asymmetries, and how data technologies make formerly illegible populations or activities visible.[10] Linnet Taylor, for one, has called for establishing a common direction in future data justice research in order to understand across social context how data can lead to discrimination, discipline, and control, as well as to acknowledge both the positive and negative possibilities of data.[11] While these conceptualizations of data justice are preliminary, they situate data within structural power relations and promote putting those understandings into practice.

DJ scholarship has developed alongside long-standing "digital justice" advocacy examining the relationship between datafication and social justice. The Detroit Digital Justice Coalition (DDJC), for instance, has developed principles of equal access to technology, participation by marginalized voices, common ownership of digital tools, and healthy communities.[12] DDJC members put these principles into practice through projects that directly engage with pressing issues: promoting equitable open data guidelines, hosting community-led Discovering Technology workshops called DiscoTechs to introduce both the impacts and possibilities of new technology, and producing zines to present communication and information rights in an accessible format.[13] The group puts shared understanding of, just access to, and control of data at the core of addressing broader social justice questions.

Bringing together EJ and DJ work, we believe understanding environmental dispossession today requires foregrounding debates over the authority of different forms of evidence and illustrating the logics of data extractivism and how data is made accessible or open, interpretable, and usable (or not). In earlier writing we provisionally defined EDJ as involving "community-based environmental data collection, public (especially online) accessibility of environmental data, and environmental data platforms supported by an open source online infrastructure—in particular, one that can be used and modified by local communities."[14] This definition prioritizes community self-determination in data collection, access, and interpretation, even if communities decide to privately hold data that represents them. For us environmental data justice provides a bridge between data justice, which emphasizes the disciplining power of data, and environmental justice, which asks whom data serves, providing a way of seeing data not just as a responsive tactic but as part of tactics for envisioning alternative futures.

Infrastructuring Environmental Data Justice through Data Together

As EDGI has continued to articulate an EDJ framework, we have asked ourselves how we can animate it in our practice. Interest in EDJ has translated into a stand-alone working group within EDGI, yet the concept is evident across our efforts. For example, EDGI members applied an EDJ framework to contextualize our own use of federal government data in reports on the first hundred days of the Trump administration and in public comments on proposed rule changes.[15] We also mobilize EDJ as we engage with novel forms of data stewardship through Data Together. Leaning on Susan Leigh Star's idea that what is most interesting about infrastructure is how it is practiced, not just what it is composed of, we elaborate on our experiment to "infrastructure" environmental data in ways that make decision-making more accessible and accountable.[16]

Data Together currently exists as a collaborative discussion space convened by EDGI; Protocol Labs, a company building open-source protocols and technologies to address how information on the internet is stored, located, and moved; and qri.io, a data science company developing dataset research tools on the distributed web.[17] Data Together's initiatives include monthly reading groups and conceptualizing open-source prototypes for decentralized archiving of web pages and datasets. The Data Together partnership grew out of conversations among DataRescue volunteers about the influence existing web technologies

Walker et al.

and internet infrastructure have on models of environmental data stewardship.[18] We have an interest in exploring whether new technologies can reframe stewardship and enable communities to hold copies of environmental data that represents them, that is used in decision-making about them, or that they otherwise care about.

One of the web technologies most of interest to Data Together is the InterPlanetary File System (IPFS), a decentralized and peer-to-peer protocol that enables storage and retrieval of resources based on their content. Whereas most forms of storage are location-addressed—the location of the resource, often represented by a URL, is used to retrieve it—content-addressed storage relies on *the content itself* to access resources. Instead of finding (or not) an air quality dataset at the URL and server to which it was published (e.g., https://www.example.com/airquality/data.csv), when that same dataset is added to an IPFS node, a copy is made available on that machine that any user can retrieve from a network of peers who have a copy. One way to retrieve a dataset on IPFS is by using your browser to visit an IPFS gateway to access that dataset using its unique signature, or cryptographic hash (e.g., https://ipfs.io/ipfs/<hash>). With IPFS the act of retrieving content becomes the same as hosting a copy. When more people access the data, its hosting becomes more distributed.

If taken up widely, this technology could restructure how information is accessed and distributed globally on the internet. However, the use of IPFS poses questions about long-term data availability. Large, infrequently used, or specialized datasets (which arguably include environmental datasets) may not be frequently replicated and as a result would be less available without concerted intervention. However, with datasets in the hands of a community of users, an environmental regulator or corporation would be unable to simply remove data from a server and cause it to disappear. Instead of the EPA's server being the only host of a key climate change dataset—making it vulnerable to the whims of the current administration—such data could be hosted in a much more public way. Further, if the data is altered by either the original publisher or a later user, the unique hash provides a durable mechanism for verifying that the intended datasets remain available. Peer retrieval potentially provides a way for the state and industry to be held accountable for providing persistent access to existing data. It does so by providing tools for verifying alterations, yet it also suggests potential infrastructures where datasets can be held in common by coordinated peers.

EDGI does not provide technical capacity for Data Together. Instead, we bring our background in EJ and commitment to EDJ to ground experiments in praxis—through environmental governance and justice use cases that connect the development of data infrastructures to the concerns emerging directly from our collaborative data-archiving efforts. As in DataRescue, in Data Together we have found tension in prototyping new data infrastructures while also preventing technologically determinist narratives from overwhelming our approach. We tend toward skepticism of grand claims that opening internet infrastructures will solve all our problems, yet we see the importance of engaging with experimental alternatives. Informed by EJ and data justice scholarship, we have been probing *how* a broader range of people can archive, *who* controls and stewards any archive, and *where* the archive is constituted.

We next reflect on challenges and unresolved tensions arising out of Data Together discussions and how we have thought about them through EDJ. Reflection and engagement provide us insight into these questions of *how*, *who*, and *where*, surfacing underlying assumptions inherent in decentralization, avoiding uncritical adoption of a participatory or community-oriented approach, and acknowledging the materiality of new data infrastructures.

The concept of decentralized (or distributed) networks owes its popularization to Paul Baran, whose work with RAND Corporation during the Cold War focused on introducing redundancy in networks so that they could survive potential nuclear attacks.[19] While the concept proved popular in the intervening years, the efficacy of these forms of decentralization has been debated, and critical scholarship has investigated how power and control still function through noncentralized protocols.[20] The discipline of geography's engagement with questions of scale provides a further provocation to the concept of decentralization, cautioning against both falling into "the local trap" of unduly privileging decentralized networks and missing how power operates both

Walker et al.

within and between scales.[21] We must continually resist assumptions that decentralized data stewardship, as proposed by Data Together, will automatically prove just.

Within Data Together we have taken up questions around coordinated archiving of datasets by decentralized peers, seeing potential alignment with forms of community-based environmental data collection. The term *community* has been evoked for a variety of forms of collective action, matching a recent participatory turn across disciplines.[22] But as with decentralization, scholars note the danger of not carefully unpacking the ways that procommunity or participatory framing avoids important questions of power within groups and across places.[23] Thus as Data Together asks how communities can hold copies of data that represent them or that are used in decision-making about them, we also keep open the question of what community *means* and how it might work in exclusionary or homogenizing ways.

Data Together engages with the relationship of decentralized archiving to existing institutions, especially in light of the role they play for the public. For instance, decentralizing archival practices could unintentionally undermine the legal requirements of the federal government and other institutions to responsibly collect and maintain the accessibility of datasets and government records. The Federal Records Act does not currently mandate the preservation of online access, meaning that records can be removed from .gov websites as long as paper copies or offline electronic iterations exist somewhere.[24] Further, federal agencies require continued funding to ensure the continuity of data collection. As a result, underfunding has been a tactic employed to prevent the creation of environmental and climate data, as seen when the Trump administration cut NASA research on greenhouse gas emissions.[25] Shifting to decentralized approaches reframes the public pressure the federal government and institutions face to reform current practices and could place an undue burden on communities and organizations that do not have the capacity to archive or collect the immense amount of data generated by the government. As Data Together continues to engage in community-oriented archiving, we need to consider the existing roles institutions play and whether and how they could advance decentralized approaches that recognize and respond to the imperfect world in which we live.

On a related note, the fact that some Data Together partners are commercial actors raises some concern in light of ongoing trends toward the privatization of knowledge.[26] Commercialized data infrastructures mean that the direction of tech development—and therefore the features of many of the current tools at our disposal—is set by private companies.[27] Commercialization is also problematic because it focuses capacity within private entities; when start-ups fail or are bought out, resources that communities were counting on may not remain available. These issues are part of important but ongoing conversations. Here we note that the issue of data loss due to its centralization in one actor (commercial or otherwise) is exactly what Data Together is meant to address.

Finally, Data Together sees promise in the coordinated distribution of datasets and other resources at the peer or node scale of the internet. Despite the shift in scale, decentralization evokes something akin to the ethereal discourses of cloud computing. The reality is that any shift in scale will have material ramifications that must be accounted for. For instance, we have seen how distributed and peer-to-peer technologies (e.g., Bitcoin's blockchain) have had significant environmental costs through electrical usage.[28] Scholarly engagement with infrastructure and material culture has prioritized the study of objects and relations that constitute the current internet, revealing a geography of fiber-optic cables, data centers, and the electromagnetic spectrum.[29] Calling attention to these "hidden" infrastructures, scholars have developed infrastructural frameworks and inversions for tackling their "taken-for-granted-ness."[30] Moving forward with Data Together, we look to explore what emerges by foregrounding the materiality of data infrastructure, especially its invisible labor and environmental costs.

As Data Together engages with questions about what it means for a broader range of people to archive, which actors have control over the archive, and the materiality of the archive, we anticipate bringing EDJ even further into the conversation. Doing so will inform our continued reflection on the role decentralization plays in infrastruc-

Walker et al.

tures that prioritize community self-determination in data collection, access, stewardship, and interpretation.

Conclusions

EDGI's data-archiving collaborations have progressed from a stance of saving data to one of reflexively engaging in experiments to create alternative social and technical infrastructures for data stewardship. The DataRescue movement raised important questions about data infrastructures and the politics of data. For us EDJ sits at the intersection of emerging DJ concerns that emphasize the disciplining effect of data surveillance and the need to "thicken" or contextualize data and EJ research and advocacy that place dispossessed communities at the center of defining and resisting injustices, but in ways that do not pathologize such communities as "damaged" and that highlight how the *lack* of, or inconsistencies in, environmental monitoring and data collection can represent injustice.[31]

Throughout this chapter we have reflected on challenges and unresolved tensions arising out of EDGI's convening of Data Together and discussed how our awareness of them has been informed by this notion of EDJ. While the election of Joe Biden in 2020 has markedly shifted the tenor of political rhetoric and action on the part of the U.S. federal government, we still live in the contours of what has been called a "post-truth" world, in which the ties between state and industry culminate in authoritarian attempts to dismiss data and facts and to downplay environmental and climate trauma. Even—especially—now we believe that the question of data must be addressed, both to recognize how data enables environmental injustice and how data could be used by communities to name and contest it.

Acknowledgments

The authors of this chapter are all part of EDGI, a fundamentally collaborative organization discussed in this commentary, whose practices and activities would not be possible without the coordinated efforts of all our members. A version of this chapter was previously published in *Geo: Geography and Environment*. We appreciate the encouraging and incisive comments from three anonymous reviewers, *Geo* editor Fiona Nash, and Jenny Goldstein. We would like to acknowledge the DataRescue organizers and volunteers, open-source contributors, and members of EDGI and Data Together who have worked collaboratively to realize the projects we describe in this chapter; it would not have been possible without our combined efforts. In addition, we thank Jeffrey Liu, Eben Pendleton, Brendan O'Brien, Liz Barry, Raymond Cha, Ed Summers, and additional reviewers for their generous feedback on earlier drafts. EDGI has received grants from Doris Duke Charitable Foundation and the David and Lucile Packard Foundation, as well as donations from individual donors.

Notes

1. Dillon et al., "Situating Data"; Dillon et al., "Environmental Data Justice"; Nost et al., *New Digital Landscape*; Rinberg et al., *Changing the Digital Climate*; Tirrell et al., "Learning in Crisis"; Vera et al., "Data Resistance"; Vera et al., "When Data Justice and Environmental Justice Meet."

2. D'Ignazio and Klein, *Data Feminism*.

3. Dillon et al., "Environmental Data Justice."

4. Dahl, "Does Secrecy Equal Security?"; Learn, "Canadian Scientists Explain."

5. Vera et al., "When Data Justice and Environmental Justice Meet."

6. Lamdan, "Lessons from DataRescue."

7. Agyeman et al., "Trends and Directions"; Altman et al., "Pollution Comes Home"; Brown, "Popular Epidemiology"; Corburn, *Street Science*; Ottinger, *Refining Expertise*; Shapiro, Roberts, and Zakariya, "Wary Alliance."

8. Allen, *Uneasy Alchemy*; Lerner, *Diamond*; Ottinger, "Constructing Empowerment."

9. Dillon et al., "Environmental Data Justice," 186.

10. Dencik, Hintz, and Cable, "Towards Data Justice?"; Dencik et al., "Exploring Data Justice"; Heeks and Renken, "Data Justice for Development"; Johnson, "From Open Data to Information Justice."

11. Taylor, "What Is Data Justice?"

12. DDJC, *Communication*; Slager, "Infrastructures of Survival."

13. DDJC and DCTP, *Recommendations for Equitable Open Data*; DDJC, "Zines."

14. Dillon et al., "Environmental Data Justice," 187.

15. Vera et al., "When Data Justice and Environmental Justice Meet."

16. Star, "Ethnography of Infrastructure"; Star and Ruhleder, "Steps towards an Ecology."

17. Data Together, https://datatogether.org.

18. Baker and Yarmey, "Data Stewardship."

19. Baran, "On Distributed Communications Networks"; Rosenzweig, "Wizards, Bureaucrats, Warriors."

20. Chun, "Crisis, Crisis, Crisis"; Galloway, "Protocol."

Walker et al.

21. Born and Purcell, "Avoiding the Local Trap"; McCarthy, "Scale, Sovereignty, and Strategy"; Swyngedouw and Heynen, "Urban Political Ecology."

22. Marres, *Material Participation*; Sieber, "Public Participation."

23. Rocheleau, Thomas-Slayter, and Wangari, *Feminist Political Ecology*; Sultana, "Community and Participation."

24. Lamdan, "Lessons from DataRescue."

25. Voosen, "Trump White House."

26. Mirowski, *Science-Mart*.

27. Eghbal, *Roads and Bridges*.

28. Lally, Kay, and Thatcher, "Computational Parasites and Hydropower."

29. Burrington, *Networks of New York*; Starosielski, *Undersea Network*; Wong and Jackson, "Wireless Visions." See also chapters 1 and 4 of this volume.

30. Star and Bowker, "How to Infrastructure."

31. Tuck, "Suspending Damage."

Bibliography

Agyeman, Julian, David Schlosberg, Luke Craven, and Caitlin Matthews. "Trends and Directions in Environmental Justice: From Inequity to Everyday Life, Community, and Just Sustainabilities." *Annual Review of Environment and Resources* 41, no. 1 (2016): 321–40.

Allen, Barbara. *Uneasy Alchemy: Citizens and Experts in Louisiana's Chemical Corridor Disputes*. Cambridge MA: MIT Press, 2003.

Altman, Rebecca Gasior, Rachel Morello-Frosch, Julia Green Brody, Ruthann Rudel, Phil Brown, and Mara Averick. "Pollution Comes Home and Gets Personal: Women's Experience of Household Chemical Exposure." *Journal of Health and Social Behavior* 49, no. 4 (December 2008): 417–35.

Baker, Karen S., and Lynn Yarmey. "Data Stewardship: Environmental Data Curation and a Web-of-Repositories." *International Journal of Digital Curation* 4, no. 2 (October 15, 2009): 12–27.

Baran, Paul. "On Distributed Communications Networks." *IEEE Transactions on Communications Systems* 12 (1964): 1–9.

Born, B., and M. Purcell. "Avoiding the Local Trap: Scale and Food Systems in Planning Research." *Journal of Planning Education and Research* 26, no. 2 (December 1, 2006): 195–207.

Brown, Phil. "Popular Epidemiology and Toxic Waste Contamination: Lay and Professional Ways of Knowing." *Journal of Health and Social Behavior* 33, no. 3 (1992): 267–81.

Burrington, Ingrid. *Networks of New York: An Illustrated Field Guide to Urban Internet Infrastructure*. Brooklyn: Melville House, 2016.

Chun, Wendy. "Crisis, Crisis, Crisis, or Sovereignty and Networks." *Theory, Culture & Society* 28 (2011): 91–112.

Corburn, Jason. *Street Science: Community Knowledge and Environmental Health Justice*. Cambridge MA: MIT Press, 2005.

Dahl, Richard. "Does Secrecy Equal Security? Limiting Access to Environmental Information." *Environmental Health Perspectives* 112, no. 2 (2004): A104.

DDJC (Detroit Digital Justice Coalition). *Communication Is a Fundamental Human Right*. Detroit: DDJC, 2009. http://detroitdjc.org/wp-content/uploads/2010/09/ddjc_1_2009.pdf.

———. "Zines." Last updated October 25, 2016. http://detroitdjc.org/zines.

DDJC and DCTP (Detroit Digital Justice Coalition and Detroit Community Technology Project). *Recommendations for Equitable Open Data*. Detroit: DDJC and DCTP, 2017. https://datajustice.github.io/report/.

Dencik, Lina, Arne Hintz, and Jonathan Cable. "Towards Data Justice? The Ambiguity of Anti-Surveillance Resistance in Political Activism." *Big Data & Society* 3, no. 2 (December 2016): 205395171667967.

Dencik, Lina, Arne Hintz, Joanna Redden, and Emiliano Treré. "Exploring Data Justice: Conceptions, Applications and Directions." *Information, Communication & Society* 22, no. 7 (June 2019): 873–81.

D'Ignazio, Catherine and Lauren F. Klein. *Data Feminism*. Cambridge MA: MIT Press, 2020.

Dillon, Lindsey, Dawn Walker, Nicholas Shapiro, Vivian Underhill, Megan Martenyi, Sara Wylie, Rebecca Lave, Michelle Murphy, Phil Brown, and EDGI. "Environmental Data Justice and the Trump Administration: Reflections from Environmental Data and Governance Initiative." *Environmental Justice* 10, no. 6 (October 2017): 186–92.

Dillon, Lindsey, Rebecca Lave, Becky Mansfield, Sara Wylie, Nicholas Shapiro, Anita Say Chan, and Michelle Murphy. "Situating Data in a Trumpian Era: The Environmental Data and Governance Initiative." *Annals of the American Association of Geographers* 109, no. 2 (January 2019): 1–11.

Eghbal, Nadia. *Roads and Bridges: The Unseen Labor behind Our Digital Infrastructure*. New York: Ford Foundation, 2016. https://www.fordfoundation.org/media/2976/roads-and-bridges-the-unseen-labor-behind-our-digital-infrastructure.pdf.

Galloway, Alexander R. "Protocol." *Theory, Culture & Society* 23, nos. 2–3 (May 2006): 317–20.

Heeks, Richard, and Jaco Renken. "Data Justice for Development: What Would It Mean?" *Information Development* 34, no. 1 (November 2016): 90–102.

Johnson, Jeffrey Alan. "From Open Data to Information Justice." *Ethics and Information Technology* 16, no. 4 (December 2014): 263–74.

Lally, Nick, Kelly Kay, and Jim Thatcher. "Computational Parasites and Hydropower: A Political Ecology of Bitcoin Mining on the Columbia River." *Environment and Planning E: Nature and Space* (August 2019): 251484861986760.

Lamdan, Sarah. "Lessons from DataRescue: The Limitations of Grassroots Climate Change Data Preservation and the Need for Federal Records Law Reform." *University of Pennsylvania Law Review* 166 (2018): 231–48.

Learn, Joshua Rapp. "Canadian Scientists Explain Exactly How Their Government Silenced Science." *Smithsonian Magazine*, January 30, 2017. https://www.smithsonianmag.com/science-nature/canadian-scientists-open-about-how-their-government-silenced-science-180961942/.

Lerner, Steve. *Diamond: A Struggle for Environmental Justice in Louisiana's Chemical Corridor*. Cambridge MA: MIT Press, 2005.

Marres, Noortje. *Material Participation: Technology, the Environment and Everyday Publics*. New York: Palgrave Macmillan, 2012.

McCarthy, James. "Scale, Sovereignty, and Strategy in Environmental Governance." *Antipode* 37 (2005): 731–53.

Mirowski, Phil. *Science-Mart: Privatizing American Science*. Cambridge MA: Harvard University Press, 2011.

Nost, Eric, Gretchen Gehrke, Aaron Lemelin, Steven Braun, Marcy Beck, Rob Brackett, Dan Allan et al. *The New Digital Landscape: How the Trump Administration Has Undermined Federal Web Infrastructures for Climate Information*. Environmental Data and Governance Initiative, July 22, 2019. https://envirodatagov.org/publication /the-new-digital-landscape-how-the-trump-administration-has-undermined-federal -web-infrastructures-for-climate-information/.

Ottinger, Gwen. "Constructing Empowerment through Interpretations of Environmental Surveillance Data." *Surveillance & Society* 8, no. 2 (December 2010): 221–34.

——— . *Refining Expertise: How Responsible Engineers Subvert Environmental Justice Challenges*. New York: NYU Press, 2013.

Rinberg, Toly, Maya Anjur-Dietrich, Marcy Beck, Andrew Bergman, Justin Derry, Lindsey Dillon, Gretchen Gehrke et al. *Changing the Digital Climate: How Climate Change Web Content Is Being Censored under the Trump Administration*. Environmental Data and Governance Initiative, January 10, 2018. https://envirodatagov.org /publication/changing-digital-climate/.

Rocheleau, Dianne, Barbara Thomas-Slayter, and Esther Wangari. *Feminist Political Ecology: Global Issues and Local Experience*. New York: Routledge, 2006.

Rosenzweig, Roy. "Wizards, Bureaucrats, Warriors, and Hackers: Writing the History of the Internet." *American Historical Review* 103, no. 5 (December 1998): 1530.

Shapiro, Nicholas, Jody Roberts, and Nasser Zakariya. "A Wary Alliance: From Enumerating the Environment to Inviting Apprehension." *Engaging Science, Technology, and Society* 3 (September 2017): 575.

Sieber, Renee. "Public Participation Geographic Information Systems: A Literature Review and Framework." *Annals of the Association of American Geographers* 96, no. 3 (September 2006): 491–507.

Star, Susan Leigh. "The Ethnography of Infrastructure." *American Behavioral Scientist* 43, no. 3 (November 1999): 377–91.

Star, Susan Leigh, and Geoffrey C. Bowker. "How to Infrastructure." In *Handbook of New Media: Social Shaping and Social Consequences of ICTs*, edited by Leah A. Lievrouw and Sonia Livingstone, 151–62. London: SAGE, 2002.

Star, Susan Leigh, and Karen Ruhleder. "Steps towards an Ecology of Infrastructure: Complex Problems in Design and Access for Large-Scale Collaborative Systems." *Proceedings of the 1994 ACM Conference on Computer Supported Cooperative Work* (1994): 253–64.

Starosielski, Nicole. *The Undersea Network*. Durham nc: Duke University Press, 2015.

Sultana, Farhana. "Community and Participation in Water Resources Management: Gendering and Naturing Development Debates from Bangladesh." *Transactions of the Institute of British Geographers* 34, no. 3 (July 2009): 346–63.

Swyngedouw, Erik, and Nik Heynen. "Urban Political Ecology, Justice and the Politics of Scale." *Antipode* 35 (2003): 898–918.

Taylor, Linnet. "What Is Data Justice? The Case for Connecting Digital Rights and Freedoms Globally." *Big Data & Society* 4, no. 2 (December 2017): 205395171773633.

Tirrell, Chris, Laura Senier, Sara Ann Wylie, Cole Alder, Grace Poudrier, Jesse DiValli, Marcy Beck, Eric Nost, Rob Brackett, and Gretchen Gehrke. "Learning in Crisis: Training Students to Monitor and Address Irresponsible Knowledge Construction by US Federal Agencies under Trump." *Engaging Science, Technology, and Society* 6 (January 2020): 81–93.

Tuck, Eve. "Suspending Damage: A Letter to Communities." *Harvard Educational Review* 79, no. 3 (2009): 409–28.

Vera, Lourdes A., Dawn Walker, Michelle Murphy, Becky Mansfield, Ladan Mohamed Siad, Jessica Ogden, and EDGI. "When Data Justice and Environmental Justice Meet: Formulating a Response to Extractive Logic through Environmental Data Justice." *Information, Communication & Society* 22, no. 7 (June 2019): 1012–28.

Vera, Lourdes A., Lindsey Dillon, Sara Wylie, Jennifer Liss Ohayon, Aaron Lemelin, Phil Brown, Christopher Sellers, Dawn Walker, and EDGI. "Data Resistance: A Social Movement Organizational Autoethnography of the Environmental Data and Governance Initiative." *Mobilization: An International Quarterly* 23, no. 4 (December 2018): 511–29.

Voosen, Paul. "Trump White House Quietly Cancels NASA Research Verifying Greenhouse Gas Cuts." *Science*, May 9, 2018. https://www.sciencemag.org/news/2018/05/trump-white-house-quietly-cancels-nasa-research-verifying-greenhouse-gas-cuts.

Wong, Richmond, and Steven Jackson. "Wireless Visions: Infrastructure, Imagination, and US Spectrum Policy." *Proceedings of the 18th ACM Conference on Computer Supported Cooperative Work & Social Computing* (2015): 105–15.

Walker et al.

Three

*Governing Data, Infrastructuring
Land and Resources*

ELEVEN

"A Poverty of Data"?

EXPORTING THE DIGITAL REVOLUTION
TO FARMERS IN THE GLOBAL SOUTH

Madeleine Fairbairn and Zenia Kish

We stand on the cusp of a data-driven revolution in global agriculture. At least, this is the shared vision within much of the international development community, where many believe that "farmers live on the edge of poverty, guided only by their history with the land and their community's traditions. Their skills and knowledge are impressive, but they suffer from a poverty of data."[1] The rapid advances in data collection, storage, and analytics that have taken place in recent years are thus seen as essential to achieving food security and other global development goals. The data revolution is attributed with the power of increasing agricultural yields, improving market access, introducing more sustainable production practices, and increasing prosperity for farmers across the Global South.[2] This chapter examines these development narratives associated with the digitalization of global agriculture, dissecting the claims used to support data-intensive solutions for the challenges faced by smallholder farmers.

In the Global North the push for more data-intensive agricultural production and value chains has been in motion for over two decades. These changes began with the widespread adoption of "precision agriculture" in the 1990s, enabled by the Global Positioning System (GPS) and other geospatial technologies, and entered a new phase in the early 2010s with increasingly affordable remote and on-farm sensors, as well as advances in cloud-based storage and big-data analytics. It is only recently, however, that a parallel effort has emerged to reimagine smallholder farming through data-driven agriculture in the Global South.[3] This effort is being led by development organizations, including major donors (e.g., the World Bank, Bill and Melinda Gates Foundation, Grameen Foundation), agricultural research institutes (e.g., members of the Consultative Group for International Agricultural Research consortium), government entities (e.g., the Dutch Ministry of Foreign Affairs, U.S. Agency for International Development [USAID]), and nonprofit nongovernmental organizations (NGOS) (e.g., Akvo, Digital Green). Private-sector companies are also important players, including those working in precision agriculture and predictive analytics (e.g., aWhere), information and advisory services (e.g., Esoko, Farmerline), and financial services (e.g., FarmDrive).

These initiatives seek to foster flows of digital data both *from* and *to* farmers. Many involve collecting data from farmers, such as farm location, crop type and seed variety, inputs applied, yields, and farm expenses; most also offer better data to farmers, including market prices, weather forecasts, soil data, and input recommendations. Frequently, data collection and dissemination go hand in hand. For example, location and operations information collected from farmers is sometimes combined with remotely sensed weather and soil data to provide farmers with recommendations for seed varieties and fertilizer application.

A growing body of scholarly research focuses on the socioeconomic and political implications of digital agriculture in the Global North.[4] This work identifies several areas in which digitalization may adversely affect farmer livelihoods, including the high cost of the technologies, limited farmer access to the data or ability to repair digital technologies, questionable protections for farmer data security and privacy, contested ownership of farm data, and further corporate concentration within the agricultural sector.[5] Though this body of research addresses many challenges posed by digital agriculture, it focuses almost exclusively on the technologies and platforms marketed to relatively large-scale farmers in wealthy countries.

Understanding the unique implications of digitalization for small farmers in the Global South requires that we also draw insights from critical research on development and humanitarian data. Efforts within international development to "bridge the digital divide" between the Global North and South have been critiqued for reproducing technocratic discourses of modernization while steering developing countries into new forms of dependency on Western epistemologies and technologies.[6] Meanwhile, the rise of "digital humanitarianism"—which draws on new forms of data to map natural disaster impacts and track refugees displaced by war, among other applications—raises questions about the privacy, identifiability, and exploitation of vulnerable populations.[7] Agricultural digitalization has very different implications for farmers in postcolonial contexts shaped by uneven exchanges of capital and expertise.

This chapter examines the consolidation of a sociotechnical imaginary that posits data-driven agriculture as crucial to modernizing farming in the Global South. Drawing from influential development agency and NGO reports on agricultural data for small farmers, as well as interviews with ag-data practitioners, we critique the narratives surrounding this push.[8] We argue that, as is commonly the case in development work, the technical tool kit supplied by data-driven agriculture can be seen as the solution to smallholder poverty only when the problem is defined in a particular way. The core problem, according to development industry publications, is that poor farmers lack the information needed to navigate today's fast-changing world—a world in which unprecedented climatic patterns and mercurial markets render farmer knowledge inadequate or chronically out-of-date. The solution to the problem thus framed is to collect more data about farmers and their crops, both for repackaging and delivery back to farmers and for sharing with other value-chain actors to multiply opportunities for profit. The depiction of farmers as having insufficient knowledge reinforces an unequal knowledge politics that dates back to colonial times, while the celebration of digital connectivity glosses over the unequal distribution of power between small farmers and the corporate actors to whom they are now connected. We argue that the sociotechnical imaginary shaping these development interventions may serve to reinforce the political-economic causes of farmer poverty.

Data Deficits: Problematizing Farmer Knowledge

To understand the revolutionary potential being attributed to digital agricultural data, we must first ask what problem it purports to solve. In the field of international development, the delineation of problems is inextricably linked to the formulation of solutions.[9] As Tania Li observes, social and environmental problems are often "rendered technical"—framed in ways that anticipate expert intervention and ignore root political-economic causes that cannot be addressed through technical fixes.[10] Among the community of international development practitioners working on data-based agricultural solutions, a single central problem is imagined: a data deficit among farmers. Farmers do not have enough information, and the information they have is not good enough.

While this problem definition seems relatively neutral, it occurs within a historical context in which purported deficiencies in farmer knowledge, technology, or abilities have often been used to justify Western interventions in the lives of rural communities globally. British colonial authorities saw their civilizing and modernizing role in the territories they occupied as extending into agriculture, which they believed could be improved with their imported knowledge and tools. James Scott cites a British provincial agricultural officer in colonial Malawi who articulated it thus: "The African has neither the training, skill, nor equipment to diagnose his soil erosion troubles nor can he plan the remedial measures, which are based on scientific knowledge, and this is where I think we rightly come in."[11] The presumed superiority of Western agricultural expertise became one of the pillars supporting colonial dispossession.

After World War II, as many former colonies gained their independence, another major agricultural modernization drive took hold with the advent of the Green Revolution, a U.S.-led push to introduce high-yielding seed varieties and associated chemical inputs to farmers in Asia and Latin America. Once again Western knowledge took center stage: the Green Revolution promulgated solutions sourced from predominantly American knowledge centers—particularly university agronomy laboratories—and circulated via American knowledge brokers, especially the Rockefeller Foundation and other philanthropies. Though often embraced by governments in former colonies pursuing national development, the Green Revolution also served strategic purposes for the United States, which saw technology transfer as a means to win Cold War allies and preempt the spread of communism among restive peasant populations.[12] This project elevated a certain kind of agricultural expertise and practice—namely, resource-intensive monocropping of cash crops—greatly increasing agricultural yields but also contributing to dependence on expensive inputs, indebtedness, over-use of water, and land loss among many poor farmers.[13] One of the goals of the Green Revolution was to modernize rural communities, transforming the "tradition-bound peasant" into a "productive, enter-prising subject."[14]

As with these previous agricultural interventions, the push for data-driven agriculture is embedded in a knowledge politics that implicitly valorizes some kinds of information (i.e., remotely sensed, technolog-ically mediated, abstract, transferable, and easily commodified) while implicitly devaluing others (i.e., informed by local history and vocab-ulary or born of physical experience).[15] These knowledge politics were made surprisingly explicit at a 2014 conference on the topic of "har-nessing the data revolution" for water and food security.[16] In his key-note lecture to an international audience of scientists, businesspeople, and policy experts, former CEO of the Gates Foundation Jeff Raikes painted a vivid picture of an African continent defined by a paucity of agricultural information and hence low productivity: "Individual farmers and the agro-dealers and extension agents who are supposed to advise them don't have the information—they don't *know* what to plant, when to plant, or how to manage soil in order to get the most out of it. Soil management practices are highly dependent on time- and location-specific characteristics, so we need near real-time and location-specific soil information to inform decision making at every level from an individual farm to a national agricultural ministry."[17] As a remedy to these knowledge deficiencies, Raikes offered up a soil-mapping project funded by the Gates Foundation, which used "mobile devices, GPS, satellite imaging, spectral lab analysis and advanced sta-tistical techniques" to survey soil across eighteen million square miles of sub-Saharan Africa.

The valorization of digital data and devaluing of existing farmer knowledge often occur in tandem. In this case the type of knowledge provided by a digital soil map becomes superior only when couched in a narrative that paints farmer knowledge as clearly deficient. That the soil data is sensed from space and compiled remotely by urban tech-nology workers did not stop Raikes from asserting its superiority as a form of "location-specific soil information." He went on to sum up the promise of digital agricultural data interventions, once again painting a picture of deep deficits in smallholder knowledge: "It's easy to imagine the 'before' and the 'after' picture, and it's stark. Now, smallholder farmers are planting whatever seed they happen to have based on roughly no information about how to conserve resources or maximize yields. In the future, based on projects that are already underway, they will plant seeds specifically adapted to the soil and water and the weather conditions they face."[18] The description of African farmers as having "roughly no information" about their own farms echoes the colonial official quoted by Scott, as does the implied need for Western intervention.[19]

An increasingly common and more nuanced version of this deficiency narrative depicts farmer knowledge as inadequate only as a result of rapidly changing environmental conditions. While farmers' knowledge about their land and crops was sufficient in the past, this variant goes, growing environmental volatility associated with climate change has created the need for continually up-to-date digital data. A USAID-funded report puts it this way: "Smallholder farmers are facing an ever-changing world: the seeds that worked for generations may not be the seeds that work today due to climate change, soil degradation, and water constraints. A farmer can no longer rely on historical calendars and generational knowledge to drive decisions about purchases, seeds to plant, and mulches and fertilizers to use. He or she needs more timely and responsive support."[20]

The coordinator of the Dutch government-funded program Geodata for Agriculture and Water put it similarly in a published interview: "Farmers still depended on the traditional signals that they picked up on from nature. But unfortunately we're having to deal with climate change. That means weather seasons are starting earlier or later, droughts are getting worse or rain is intensifying. Essentially weather has become much more unpredictable. . . . Satellites generate much more precise, location-based, and therefore reliable, data."[21] In a twist on development institutions' long-standing tendency to dismiss peasant knowledge as backward and resistant to change, it is here framed as unreliable only because it has been outpaced by external change.

Fairbairn and Kish

The "generational knowledge" and "traditional signals" of farmers, it is implied, are relatively static, whereas digital data is valorized as nimbler and more precise. In addition to climate change, digital development reports also highlight population growth, market volatility, and unpredictable crop epidemics as causes of instability that progressively render farmer knowledge obsolete. The "small data" of farmers, the logic goes, is too slow, fragmented, and imprecise to allow them to keep up with a rapidly changing world.

Agricultural "Data Ecosystems": Connectivity as Solution

The data deficit narrative we have described supports a sociotechnical imaginary that simplifies the complex social and economic difficulties of farmers by examining them through a narrow technical lens. This framing sets up an equally technical solution: only through increased flows of digital data can farmers adapt to changing market and environmental conditions. Like the problematization of farmer knowledge, this solution framing reproduces modernization discourses that stretch back to colonial encounters and have been revived with the enthusiastic embrace of information communication technologies for development.[22] Of particular relevance here is the long-standing modernist faith in connective technologies—from railroads to logistics infrastructure to mobile phones—as means to stimulate "backward" peripheral economies and foster cultures of innovation within them.[23] Connectivity has long been depicted as a conduit for the diffusion of development from rich to poor countries, with little recognition of how such articulations also produced underdevelopment by facilitating the extraction of value.[24] This belief in connectivity can be seen currently in relation to smallholder agriculture, where the increasing collection, integration, and dissemination of digital data is depicted as essential to boost productivity and profit among poor farmers across the developing world.

The need for greater, more connected flows of digital data on agriculture in the Global South is framed largely in terms of the benefits that will accrue to smallholders. But this rarely means that farmers are being given direct access to the data; generally, expert mediation is assumed to be required to transform agricultural data into information that farmers can comprehend and use. "The true data revolution does not simply lie in floodgates of data suddenly opening up," writes André Laperrière, executive director of the Global Open Data for Agriculture and Nutrition (GODAN) initiative, in a discussion paper titled simply *Data Revolution for Agriculture*; "it lies in concrete applications that make sense and use of it."[25] The interventions of organizations such as USAID or the Gates Foundation are seen as providing necessary technological brokerage for farmers who would otherwise be at a loss to translate raw data into usable, potentially profitable information. While these development community or private-sector actors remain at the center of data flows to and from smallholder farmers, digital data is nonetheless framed as a means to uplift small farmers, producing not just agronomic improvements but also farmer empowerment. In the same GODAN paper, Laperrière asserts that "knowledge is no more the privilege for a few, but a right for everyone," and goes on to quote a publication by the McKinsey consulting firm calling the data revolution "globalisation for the little guy."[26] Datafication is framed as a route not just to modernization but to the democratization of knowledge.

Yet data flows and the benefits they promise extend far beyond farmers. Smallholder farmers are visualized as part of a broader "data ecosystem" that includes other agricultural value-chain actors (e.g., input vendors, wholesalers, processors), service providers (e.g., extension agents, agricultural lenders), and even governance actors (e.g., various levels of government, NGOs).[27] Freer flows of data among these various actors are seen as generating positive and synergistic benefits for everyone involved. As the *Data Revolution for Agriculture* report puts it, "Each of these actors is interested in information about production, functioning of the value chain and availability of services and governance. The more and better the information that is available about the smallholders' ecosystem and its functioning, the better the different actors can fulfil their role, ultimately strengthening smallholders' food and nutrition security."[28] This sociotechnical imaginary highlights data connectivity and related technologies as capable of unlocking economic growth not only for smallholders but for all value-chain actors.[29] Farmers thus become a single node within the

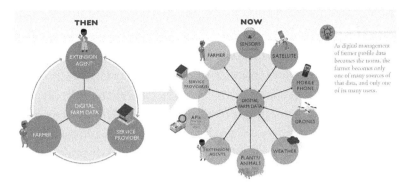

19. USAID graphic illustrating the digital data ecosystem of the smallholder farmer. USAID.

significantly larger data ecosystem that serves many interests beyond the farmers themselves.

In line with the ethos of big data, the aim is to expand the volume, variety, and velocity of data available across the farm data ecosystem. As illustrated in a USAID infographic on the benefits of digital farmer profiles (fig. 19), farm data supplied by farmers is combined with data from many other sources. The ultimate goal, according to USAID, is to minimize reliance on farmers themselves to provide this data: "New ways of collecting and aggregating data and applying analytics—such as predictive, prescriptive, and cognitive analytics—can reduce the amount of direct input needed from the farmer. Data analytics is a game-changer and is being used to create 'new data' from existing data."[30] Optimizing value chain profitability and efficiency begins with rendering more comprehensive digital portraits of small farms by collecting as many data points as possible, with or without farmers' direct participation. Data and its supporting technologies are given center stage as solutions for the challenges experienced by farmers, while obfuscating the socioeconomic and political systems that shape farmer realities.

Data Revolution: Political Economic Implications

The problem and solution narratives we have described are important because they risk reproducing the unequal social relations—particularly relations of dependence on purchased inputs, credit financing, and the corporate value-chain actors that provide them—that are largely to blame for smallholder poverty in the first place. Here the "data revolution" could learn from the history of the original Green Revolution, in which triumphant narratives of agricultural progress through North-South knowledge transfer were often accompanied by the growing economic vulnerability of small farmers.[31] The shift to more capital-intensive modes of farming fostered under the Green Revolution made farmers more dependent on corporate vendors of agricultural machinery and inputs. In replacing saved seeds with purchased hybrids and animal manure with purchased fertilizer, the Green Revolution expanded markets for northern agribusiness while contributing to farmer indebtedness across Asia and Latin America.[32] The seemingly apolitical sociotechnical imaginaries associated with the Green Revolution masked political economic effects that were far from neutral.[33]

In keeping with these entangled histories of technological modernization and capitalist consolidation, today's data-driven agricultural interventions serve to intensify relationships between small farmers and corporate actors. First, on a most basic level, the goal of many data-driven agricultural interventions is to give farmers the information they need to optimize their applications of fertilizer and other purchased inputs.[34] This goal is highly compatible with the kind of productivist agriculture that sustains corporate agri-input suppliers. In the agricultural development arena, this productivist vision can be seen at work in the drive for "sustainable intensification," a term for increasing yields without bringing more land under production. It has been embraced widely among development institutions and also increasingly among agribusinesses and industry associations, such as the International Fertilizer Industry Association and Agricultural Biotechnology Council.[35] In the Global North, precision agriculture has long been promoted as a means to reduce overall agricultural input usage by revealing parts of fields in which farmers are currently over-applying fertilizers or pesticides. In the Global South, however, baseline levels of input use are generally much lower, making an upward trajectory of input use far more likely.

Second, the datafication of agriculture frequently goes hand in hand with efforts to link smallholders to financial service providers. In this way data may serve to reconfigure how farmers interact with

Fairbairn and Kish

and become embedded in financial markets. One prominent example of this is index-based agricultural insurance. Unlike traditional crop insurance, which makes indemnity payments to farmers based on their demonstrable crop losses, index insurance payouts are based on environmental indices (e.g., rainfall, temperature) that are expected to correlate with crop loss.[36] Because of this unusual model, index insurance is quite cheap, and development organizations, including the World Bank and UN International Fund for Agricultural Development, have begun touting it as a promising agricultural development tool. Yet, as Leigh Johnson observes, in marketing index insurance to small farmers as a means of reducing their *agricultural risk*, development institutions encourage them to take on *financial risk*, thereby transforming agricultural producers into financial consumers.[37] Remotely sensed weather data, as an independent consultant we interviewed underscored, is *not* an infallible proxy for crop performance. Even if satellite data could determine rainfall in a given region with perfect accuracy (it cannot), crop performance would still depend on other variables, such as soil type, farmer skill, and topography. For farmers, therefore, buying index insurance essentially means speculating on the accuracy of remotely sensed data and its derived indices.

Also in the realm of financial services, many ag-data projects aim to increase farmer access to credit financing. Agronomic data such as yield predictions, field size, sales records, and landownership are increasingly being used as a basis for alternative credit scoring, which allows lenders to assess the creditworthiness of farmers even if they lack bank accounts and other traditional financial records. The Grameen Foundation, which pioneered microlending for the poor, is one of the major donor organizations working at the intersection of farm data and farm financing; Grameen-backed projects in Uganda (Opportunity Bank Partnership) and Colombia (Agricultural Risk Evaluation Tool) collect digital data on farmers as a basis for establishing creditworthiness.[38] Similarly, Nairobi-based start-up FarmDrive combines farmer-provided data (including demographic information, farm location, soil type, and economic data on expenses, revenues, and yields) with remotely sensed data on weather and soils, then algorithmically converts this data into farmer credit profiles.[39] In Ghana alone several companies (SyeComp, Esoko, and BigData Ghana) are working on some kind of alternative credit-scoring initiative. Although credit is a much-needed input among smallholder farmers, microloans can also be fertile ground for predatory lending practices and gendered repayment burdens, and therefore any effort to extend credit to small farmers must consider the economic vulnerability that comes with debt.[40]

Third, ag-data initiatives often serve to increase the visibility of small farmers to corporate agribusiness. One ag-tech start-up, for instance, integrates remotely sensed data with farmer-supplied data to provide farmers free access to a platform that allows them to visualize their fields and manage their production. However, in an interview, one of the company executives explained the trade-off required to provide this free data service to farmers: the company also has a contract with a major international agribusiness to map cropped areas of wheat and corn across parts of Africa to improve the agribusiness's field force effectiveness—its ability to sell inputs directly to farmers. The digital extension services provided to smallholders are therefore made possible by selling their data to a corporation whose profits depend, at least in part, on converting those farmers into paying customers. While the connectivity narrative celebrates increased data flows for their ability to boost profits across the "data ecosystem," the vast power asymmetry between a small African farmer and a multi-billion-dollar seed and chemical company suggests the need for deeper consideration of the possible implications of sharing farmer data. As Scott observes, increased visibility to powerful actors can have negative or even dangerous consequences for peasant communities.[41]

Agribusinesses are also developing their own platforms for gathering data on smallholders. Syngenta's Farmforce, for instance, is a software tool that allows for the use of mobile devices to track the activities of small farmers. Its website advertises the tool as helping small-scale farmers access formal markets by increasing the traceability of their crops and facilitating compliance with food safety and other standards. However, the website also makes clear that a major market for the software will be managers of outgrower schemes struggling to monitor and discipline—in the Foucauldian sense—large numbers of small farmers who produce for them under contract.[42] As we discussed earlier, the ag-data revolution privileges data that is visible to and transferable among a broad range of parties. Here we see one reason why: it allows for monitoring, which can be used to enforce the demands of formal and global markets.

Fourth and finally, ag-data services aimed at farmers are sometimes bundled with other corporate products, such as the cell phone services offered by a mobile network operator or inputs supplied by an agribusiness. In this scenario farmers who wish to receive data must also sign up for some other product or service. This approach is in line with a broader trend toward public-private partnerships as development becomes more data-centric. Linnet Taylor and Dennis Broeders argue that the use of big data in development initiatives leads to a "power shift from the traditional collector and user of statistics—the state—to a messier, more distributed landscape of governance where power accrues to those who hold the most data," with the result that corporations are becoming ever more important development actors.[43] Many of the NGO and donor representatives we interviewed described strategic partnerships with private companies as the best (or only) route to ensuring the long-term success of their digital extension services. Out of such bundling arrangements, the corporate actor gets increased market penetration by having its services promoted to farmers directly by trusted nonprofit or government organizations, while the nonprofit or government organizations get assurance that their extension approach will last beyond the length of the grant funding because of being integrated into a revenue-generating business model. Whether the farmer is best served by such strategic collaborations is another question.

While greater access to fertilizer, credit, or mobile networks may be a boon to farmers in many cases, to assume that greater market penetration is a positive development in its own right is to be willfully blind to the history of agricultural development. Before embracing the transformative potential of agricultural data, development practitioners should ask themselves why it is that agribusinesses, banks, or mobile network operators are so willing to collaborate with them and what relations of dependency they may be helping to foster within rural communities.

Conclusions

While development practitioners and private sector actors tout the "revolutionary" potential of data-driven agriculture for small farmers in the Global South, their narratives betray a knowledge politics that echoes the past. In this farmers are defined—as they have been since colonization—through a rhetoric of deficits: of technology, of knowledge, of modernity, of adaptability to a changing world. Enhanced collection and dissemination of farmer data is posed as a solution, sidestepping any critical accounting of how the structural residues of colonialism, agricultural intensification, and neoliberal development policies have played a role in producing farmer vulnerability in the first place. Instead, the sociotechnical imaginary underpinning data-driven agriculture envisions a pervasive digital integration in which farms of all scales will be increasingly digitized, their data will be more visible (and profitable) to agribusinesses and other value-chain actors, farmers will participate more fully in financial and other privately supplied services, and Western narratives of progress through science remain unchallenged. This vision is attractive for both agribusinesses and nonprofits because it promises both sets of actors coveted public-private partnerships that will potentially expand access to, respectively, customers and funding.[44] Yet bereft of any substantial consideration of the causes of small farmer poverty, it also risks accelerating agricultural consolidation.

Under the right social conditions, digital data is potentially compatible with more emancipatory agricultural agendas, including agroecological production practices and small farmer cooperatives.[45] However, rather than presuming that farmers are plagued first and foremost by data deficits, development agencies must *begin* by engaging in rigorous consultation with the farmers they hope to benefit, asking what problems *the farmers* wish to see solved and using this to guide the digital solutions developed. Likewise, rather than assuming the inherent benefits of data sharing among all ecosystem actors, development actors should empower farmers to decide which corporate, governmental, and other actors should have access to their data and which should not. Truly farmer-led initiatives to collect and share data could

Fairbairn and Kish

avoid predatory forms of monetization and offer more just development opportunities.

Notes

1. Lobell, "Big Data Has Transformed Agriculture."

2. See, for example, CTA, *Data Revolution for Agriculture*; WEF, "Innovation with a Purpose."

3. Here we use the term *smallholder* because it is the one used most by the community of development practitioners and private-sector companies we are researching. However, the term is problematic because it implies landownership, or at least stable tenure, excluding large numbers of landless farmers and farm laborers.

4. To name just a few, Bronson and Knezevic, "Big Data"; Carolan, "Agro-digital Governance"; Carolan, "Publicising Food"; Fraser, "Land Grab/Data Grab"; Wolf and Wood, "Precision Farming."

5. Rotz et al., "Politics of Digital Agricultural Technologies."

6. Hilbert, "Big Data for Development"; Ouma, Stenmanns, and Verne, "African Economies"; Pieterse, "Digital Capitalism and Development"; Stevenson, "Digital Divide"; Thompson, "Discourse."

7. Taylor, "Data Subjects or Data Citizens?"; Taylor and Broeders, "In the Name of Development."

8. We follow Jasanoff's definition of sociotechnical imaginaries as "collectively held, institutionally stabilized, and publicly performed visions of desirable futures, animated by shared understandings of forms of social life and social order attainable through, and supportive of, advances in science and technology." Jasanoff, "Future Imperfect," 4. Between 2016 and 2018 we conducted thirty-nine interviews with people working in a range of organizations focused on bringing digital solutions to farmers in the Global South, including private companies, nonprofits, development agencies, and the public sector. Though the quotes in this paper are drawn primarily from publicly available documents, these interviews inform the overall analysis.

9. Ferguson, *Anti-Politics Machine*; Li, *Will to Improve*.

10. Li, *Will to Improve*, 123.

11. Scott, *Seeing like a State*, 226.

12. Cullather, *Hungry World*.

13. Griffin, *Political Economy*; Patel, "Long Green Revolution."

14. Nally and Taylor, "Politics of Self Help," 51.

15. Elwood and Leszczynski define "knowledge politics" as "the use of particular information content, forms of representation or ways of analysing and manipulating information to try to establish the authority or legitimacy of knowledge claims." "New Spatial Media," 544.

16. The annual conference of the Robert B. Daugherty Water for Food Global Institute of the University of Nebraska. That year the conference was held in Seattle, in partnership with the Gates Foundation, and was titled "Harnessing the Data Revolution:

Ensuring Water and Food Security from Field to Global Scales." Notably, the conference received funding from such major agribusinesses as Monsanto, Syngenta, Dow, and DuPont.

17. Raikes, "Plenary Address," 40:53.

18. Raikes, "Plenary Address," 42:30.

19. That Raikes echoed the language of the Green Revolution is less surprising given that the Gates Foundation explicitly seeks to carry on its legacy through participation in the major agricultural development project Alliance for a Green Revolution in Africa.

20. Gray et al., *Digital Farmer Profiles*, 3.

21. CTA, "Precision Agriculture," 5.

22. Díaz Andrade and Urquhart, "Unveiling the Modernity Bias."

23. Ouma, Stenmanns, and Verne, "African Economies."

24. Carmody, "Informationalization of Poverty."

25. CTA, *Data Revolution for Agriculture*, ii.

26. CTA, *Data Revolution for Agriculture*, ii.

27. CTA, *Data Revolution for Agriculture*.

28. CTA, *Data Revolution for Agriculture*, 8.

29. WEF, *Innovation with a Purpose*.

30. Gray et al., *Digital Farmer Profiles*, 2.

31. Patel, "Long Green Revolution."

32. Cullather, *Hungry World*.

33. Smith, "Imaginaries of Development."

34. Maru et al., "Digital and Data-Driven Agriculture."

35. Cook et al., *Sustainable Intensification Revisited*.

36. Isakson, "Derivatives for Development?"

37. Johnson, "Index Insurance."

38. Grameen Foundation, "Connect Farmers to Finance."

39. Burwood-Taylor, "FarmDrive Raises Funding."

40. Roy, *Poverty Capital*.

41. Scott, *Seeing like a State*.

42. Farmforce, "Advanced Traceability," https://farmforce.com/.

43. Taylor and Broeders, "In the Name of Development," 229.

44. Pieterse, "Digital Capitalism and Development."

45. Carolan, "Publicising Food"; Rotz et al., "Politics of Digital Agricultural Technologies."

Bibliography

Bronson, Kelly, and Irena Knezevic. "Big Data in Food and Agriculture." *Big Data & Society* 3, no. 1 (June 2016): 1–5.

Burwood-Taylor, Louisa. "FarmDrive Raises Funding to Help Africa's Smallholder Farmers Get Finance with Credit Scoring Algorithm." *AgFunder News*, February 16, 2017. https://agfundernews.com/farmdrive-raises-funding-to-help-africas-smallholder -farmers-get-finance-with-credit-scoring-algorithm.html.

Carmody, Pádraig. "The Informationalization of Poverty in Africa? Mobile Phones and

Economic Structure." *Information Technologies and International Development* 8, no. 3 (2012): 1–17.

Carolan, Michael. "Agro-digital Governance and Life Itself: Food Politics at the Intersection of Code and Affect." *Sociologia Ruralis* 57, no. S1 (2016): 816–35.

———. "Publicising Food: Big Data, Precision Agriculture, and Co-experimental Techniques of Addition." *Sociologia Ruralis* 57, no. 2 (2017): 135–54.

Cook, Seth, Laura Silici, Barbara Adolph, and Sarah Walker. *Sustainable Intensification Revisited.* IIED Issue Paper. London: International Institute for Environment and Development (IIED), 2015. https://pubs.iied.org/pdfs/14651IIED.pdf.

CTA (Technical Centre for Agricultural and Rural Cooperation). *Data Revolution for Agriculture.* CTA Discussion Paper. Wageningen, Netherlands: CTA, 2015. https://brusselsbriefings.files.wordpress.com/2015/03/1937_pdf.pdf.

———. "Precision Agriculture for Smallholder Farmers." *ICT Update* 86 (February 2018). https://ictupdate.cta.int/en/issues/86-precision-agriculture-for-smallholder-farmers.

Cullather, Nick. *The Hungry World: America's Cold War Battle against Poverty in Asia.* Cambridge MA: Harvard University Press, 2013.

Díaz Andrade, Antonio, and Cathy Urquhart. "Unveiling the Modernity Bias: A Critical Examination of the Politics of ICT4D." *Information Technology for Development* 18, no. 4 (2012): 281–92.

Elwood, Sarah, and Agnieszka Leszczynski. "New Spatial Media, New Knowledge Politics." *Transactions of the Institute of British Geographers* 38, no. 4 (2013): 544–59.

Ferguson, James. *The Anti-Politics Machine: Development, Depoliticization, and Bureaucratic Power in Lesotho.* Minneapolis: University of Minnesota Press, 1994.

Fraser, Alastair. "Land Grab/Data Grab: Precision Agriculture and its New Horizons." *Journal of Peasant Studies* 46, no. 5 (2019): 893–912.

Grameen Foundation. "Connect Farmers to Finance." Accessed December 8, 2021. https://grameenfoundation.org/what-we-do/agriculture/connect-farmers-finance#main-content.

Gray, Bobbi, Lee Babcock, Leo Tobias, Mona McCord, Ana Herrera, Cecil Osei, and Ramiro Cadavid. *Digital Farmer Profiles: Reimagining Smallholder Agriculture.* Feed the Future Initiative. Washington DC: USAID, 2018. https://www.usaid.gov/sites/default/files/documents/15396/Data_Driven_Agriculture_Farmer_Profile.pdf.

Griffin, Keith. *The Political Economy of Agrarian Change: An Essay on the Green Revolution.* London: Macmillan, 1979.

Hilbert, Martin. "Big Data for Development: A Review of Promises and Challenges." *Development Policy Review* 34, no.1 (2016): 135–74.

Isakson, Ryan. "Derivatives for Development? Small-Farmer Vulnerability and the Financialization of Climate Risk Management." *Journal of Agrarian Change* 15, no. 4 (2015): 569–80.

Jasanoff, Sheila. "Future Imperfect: Science, Technology, and the Imaginations of Modernity." In *Dreamscapes of Modernity: Sociotechnical Imaginaries and the Fabrication of Power*, edited by Sheila Jasanoff and Sang-Hyun Kim, 1–33. Chicago: University of Chicago Press, 2015.

Johnson, Leigh. "Index Insurance and the Articulation of Risk-Bearing Subjects." *Environment and Planning A* 45, no.11 (2013): 2663–81.

Li, Tania. *The Will to Improve: Governmentality, Development, and the Practice of Politics.* Durham NC: Duke University Press, 2007.

Lobell, David. "Big Data Has Transformed Agriculture—in Some Places, Anyway." *Observations* (blog). *Scientific American*, April 18, 2019. https://blogs.scientificamerican .com/observations/big-data-has-transformed-agriculture-in-some-places-anyway/.

Maru, Ajit, Dan Berne, Jeremy de Beer, Peter G. Ballantyne, Valeria Pesce, Stephen Kalyesubula, Nicolene Fourie, Chris Addison, Anneliza Collett, and Juanita Chavez. "Digital and Data-Driven Agriculture: Harnessing the Power of Data for Smallholders." Rome: Global Forum on Agricultural Research and Innovation, 2018. https://hdl .handle.net/10568/92477.

Nally, David, and Stephen Taylor. "The Politics of Self-Help: The Rockefeller Foundation, Philanthropy and the 'Long' Green Revolution." *Political Geography* 49 (2015): 51–63.

Ouma, Stefan, Julian Stenmanns, and Julia Verne. "African Economies: Simply Connect? Problematizing the Discourse on Connectivity in Logistics and Communication." In *Digital Economies at Global Margins*, edited by Mark Graham, 341–63. Cambridge MA: MIT Press, 2019.

Patel, Raj. "The Long Green Revolution." *Journal of Peasant Studies* 40, no. 1 (2013): 1–63.

Pieterse, Jan Nederveen. "Digital Capitalism and Development: The Unbearable Lightness of ICT 4D." In *Emerging Digital Spaces in Contemporary Society: Properties of Technology*, edited by Phillip Kalantzis-Cope and Karim Gherab-Martín, 305–23. New York: Palgrave Macmillan, 2010.

Raikes, Jeff. "Plenary Address." Water for Food Global Conference, Seattle, October 20, 2014. https://youtu.be/GwIrVTyE0g8.

Rotz, Sarah, Emily Duncan, Matthew Small, Janos Botschner, Rozita Dara, Ian Mosby, Mark Reed, and Evan Fraser. "The Politics of Digital Agricultural Technologies: A Preliminary Review." *Sociologia Ruralis* 59, no. 2 (2019): 203–29.

Roy, Ananya. *Poverty Capital: Microfinance and the Making of Development.* New York: Routledge, 2010.

Scott, James. *Seeing like a State: How Certain Schemes to Improve the Human Condition Have Failed.* New Haven CT: Yale University Press, 1998.

Smith, Elta. "Imaginaries of Development: The Rockefeller Foundation and Rice Research." *Science as Culture* 18, no. 4 (2009): 461–82.

Stevenson, Siobhan. "Digital Divide: A Discursive Move Away from the Real Inequities." *Information Society* 25 (2009): 1–22.

Taylor, Linnet. "Data Subjects or Data Citizens? Addressing the Global Regulatory Challenge of Big Data." In *Information, Freedom and Property: The Philosophy of Law Meets the Philosophy of Technology*, edited by Mireille Hildebrandt and Bibi van den Berg, 81–104. New York: Routledge, 2016.

Taylor, Linnet, and Dennis Broeders. "In the Name of Development: Power, Profit and the Datafication of the Global South." *Geoforum* 64 (2015): 229–37.

Thompson, Mark. "Discourse, 'Development' and the 'Digital Divide': ICT and the World Bank." *Review of African Political Economy* 31, no. 99 (2004): 103–23.

WEF (World Economic Forum). *Innovation with a Purpose: The Role of Technology Innovation in Accelerating Food Systems Innovation.* Geneva, Switzerland: WEF, 2018. http://www3.weforum.org/docs/WEF_Innovation_with_a_Purpose_VF-reduced.pdf.

Wolf, Steven, and Spencer Wood. "Precision Farming: Environmental Legitimation, Commodification of Information, and Industrial Coordination." *Rural Sociology* 62, no. 2 (1997): 180–206.

··

TWELVE

Illicit Digital Environments

MONITORING AND SURVEILLING ENVIRONMENTAL
CRIME IN SOUTHEAST ASIA

Hilary O. Faxon and Jenny Goldstein

Big data, environmental sensors, online platforms and databases, artificial intelligence, and other Smart Earth technologies have caught on in the Global North as a way to better know, and therefore solve, environmental problems.[1] At a time of ongoing environmental degradation and amid rapid digital evolution, the promise of more openness and transparency offered by these technologies provides an opportunity to harness data infrastructures for environmental governance. Environmental governance increasingly involves diverse actors aligned in new and often unpredictable ways.[2] Southeast Asia, for example, hosts an especially diverse set of hybrid governance regimes due to mixed political and economic systems across and within nation-states, as well as the varying types of ecosystems in the region and the actors involved in their management. Hybrid governance partnerships—between and among states, private companies, communities, and civil society organizations—tend to espouse democratic ideals such as participation, often without recognizing the ways that these collaborations remain subordinate to illiberal, authoritarian-leaning governments and top-down market forces in Southeast Asia.[3] Yet as sham elections in Cambodia in 2018 and military coups in Thailand in 2014 and Myanmar in 2021 have demonstrated, regional efforts toward development and democracy are repeatedly challenged by resurgent authoritarianism. Within these political contexts, and due in part to the proliferation of data infrastructures accessible to actors outside of state institutions, hybrid governance regimes both drive and are driven by digital tools that can be used to monitor resources from within and outside of nation-state borders.

Southeast Asia, a region with major biodiversity hot spots and extensive tropical forests, as well as widespread internet and smartphone use, is a particularly rich place for understanding the effects of data infrastructures on environmental governance in the Global South.[4] Across most countries in the region, new data infrastructures are increasingly used to monitor illicit environmental activities such as small-scale logging, wildlife poaching and exotic species trading, unregulated fishing, and labor exploitation in the mining industries. New software, platforms, and smartphone apps are intended to provide information transparency in order to promote more equitable access to and sustained use of natural resources. But promises of better management through data sharing and collaboration raise questions of *how* data infrastructures surveil illicit activities and thus contribute to environmental governance. For instance, in the absence of robust enforcement systems, does digital environmental monitoring empower citizens to hold states accountable for protecting natural resources, or might it actually enable further exploitation of resources and the people who depend on them? Does the "view from nowhere" offered by satellite images reinforce existing imaginaries to promote top-down decision-making or provide actionable information for more inclusive management of inaccessible places?[5]

Despite optimism that new data infrastructures for environmental monitoring will lead to stronger environmental conservation and more equitable resource use and access throughout the tropics, research is beginning to suggest otherwise.[6] Most digital platforms in use today are based in North America and Europe, and the financial benefits of "platform capitalism" accrue to these corporations.[7] Furthermore, the normative orientations, legal frameworks, and underlying assumptions in digital platforms correspond to an individual rights–based approach that is often inadequate for addressing underlying questions of data justice, particularly in the Global South.[8] Research from South Africa, for example, has shown that the efficacy of grassroots data activism is limited by broader power relations and systems of spatial injustice.[9] Without taking context into consideration, data infrastructures for environmental monitoring in the Global South could reinforce existing hierarchies of expertise by responding to foreign goals while supplying data of little relevance to local contexts.[10] Moreover, digital tools that enable monitoring efforts are often the same ones behind the global expansion of digital surveillance—by states and private companies—of citizens and clients.[11] The threat is particularly acute in a region neighboring China, a global leader in not only developing and deploying, but also exporting, digital systems for surveillance and control.[12] Civic monitoring presents an opportunity for democratic participation, but transforming better data into more sustainable or equitable governance can be constrained by technical, political, and legal dynamics in semiauthoritarian nations with historic and ongoing patterns of surveillance and state violence.

Nonstate and extrastate actors in environmental governance are especially concerned with monitoring illicit activities, similar to the illicit land transactions that Beth Tellman and colleagues describe as "exchanges that are not allowed or permitted by formal or informal rules and norms that govern social interactions, which are enforced through various societal mechanisms or institutions."[13] Those engaging in illicit environmental activities intentionally seek to keep them hidden from public view to avoid detection, and these activities often go unrecorded in official records and result in ecological change that would not otherwise take place.[14] Normative explanations of illicit environmental activities in Southeast Asia assume that states fail to enforce regulations that would otherwise prevent such activities from occurring. Any state involvement in illicit transactions is typically deemed corruption; eliminating state corruption is understood to be a moral imperative in the service of good governance. Yet government involvement in illicit activities can also be a structural feature of functioning states. While most Southeast Asian countries have limited abilities to obtain state revenue through taxes, they maintain a strong ability to finance illicit resource extraction to facilitate institution building and economic accumulation for political elites.[15] This punctures the narrative that states in Southeast Asia lack "capacity" for state building and instead highlights how strategic, intentional support of illicit activities often provides substantial under-the-radar moneys for state coffers.

The ability of diverse actors operating both in situ and remotely to track illicit activity raises questions about environmental surveillance—pervasive monitoring coupled with force or punitive measures—and about how the detection of environmental crimes can lead to their prevention, disruption, or maintenance. More precise real-time detection of such activities does not, and perhaps should not, lead directly

Faxon and Goldstein

to more heavy-handed enforcement of existing laws or the drafting of stronger regulations. As Ben Neimark comments, "In countries where enforcement-heavy conservation strategies take precedence, the rural poor and frontline activists, rather than those truly at fault, often pay the price."[16] What is more, shifting laws and power relations surrounding environmental resources change the definitions of what is illicit and who is complicit and risk misclassifying as criminals those with lives and livelihoods at stake.[17]

This chapter explores how digital initiatives to monitor environments for illicit and extralegal activities are taking shape in Southeast Asia, where in many countries state actors are involved in or enriched by these activities, even as surveillance has been and continues to be a tool of government repression.[18] We argue that while greater digital connectivity enables new forms of environmental monitoring, the impacts of this monitoring vary across governance regimes and cultural context. Political and technological histories—particularly those concerning the relationship among citizens, the state, and information—shape the forms and potential of digital environmental monitoring efforts. In the next section we explore the political ecology of data in Southeast Asia, describing how data infrastructures are being produced and used under a diversity of governance regimes to track two major threats to forest and animal species conservation: wildlife poaching and logging. Then, using Myanmar as a case study, we ask what happens when data infrastructures for environmental monitoring are introduced in an authoritarian surveillance state experimenting with liberal democracy. We map modes of digital environmental monitoring that emerged during Myanmar's democratic reform period (2012–21) and then highlight an online platform focused on the jade sector in order to analyze common challenges in building and using data infrastructures to hold the state accountable and advocate for more just and equitable resource governance.

Digital Environmental Monitoring of Illicit Activities in Southeast Asia

New data infrastructures provide innovative tools to track the prevalence and patterns of illicit activity, though they can also contribute to perpetuating environmental crime. Extralegal environmental activities in Southeast Asia are extensive and include small-scale gold mining in Indonesia and the Philippines; opium production in the Golden Triangle, a cross-border region of Thailand, Myanmar, and Laos; large cat species poaching in Thailand, Laos, Indonesia, and Malaysia; tortoise poaching in the Coral Triangle between Indonesia, Malaysia, and the Philippines; and timber harvesting and trafficking throughout forests in Southeast Asia. Timber is a particularly extensive natural resource extracted within illicit economies; a 2016 report by INTERPOL and the UN Environment Programme states, for instance, that 73 percent of all timber harvested in Indonesia was logged illegally.[19] Categorizing these illicit activities can be tricky: governments sometimes make otherwise illicit activities legal for their own benefit, while marginalized actors, particularly the rural poor, are forced to undertake small-scale illicit activities just to maintain their livelihoods.

The predominant method of detecting illicit environmental activities over the past few decades has been remote sensing, or what Tellman and colleagues call "pixel-based approaches" to observe unexpected spatiotemporal changes in land use cover through satellite-derived data. In some cases cadastral maps, ethnographic interviews, and media reports have been used to corroborate and establish causality of land use changes detected remotely.[20] Platforms like the World Resources Institute's Global Forest Watch (GFW) have made remote sensing–derived forest cover maps available to the public in the absence of accessible, authoritative, and transparent forest maps from state agencies in countries including Malaysia and Indonesia.[21] Building custom data layers on top of existing datasets available through GFW, civil society organizations such as Hutanwatch have used the site's satellite imagery and geographic information system (GIS) data points to identify instances of illegal deforestation, validate the occurrences on the ground, and publish stories in local and national news outlets to "name and shame" the culprits. In some instances the media and ensuing public outcries for enforcement have led state forest agencies to expand plans for forest conservation.[22]

Yet remote sensing is less effective at detecting small-scale activities that lead to ecosystem degradation, such as wildlife poaching or

Faxon and Goldstein

the selective logging that constitutes most of the illicit timber harvesting in Southeast Asia.[23] In recent years environmental data infrastructures have increasingly been designed with the explicit aim of monitoring small-scale deforestation on the ground throughout the region's tropical forests, albeit with different objectives depending on the actors involved.

For instance, on the Indonesian island of Sumatra, a conservationist group from California installed a "treetop surveillance system" built from recycled mobile phones and artificial intelligence software that listens for audio signs of small-scale illegal logging. The phones capture sounds of chain saws and logging trucks, which are analyzed in real time by the software and used to send alerts to local rangers. And in Cambodia, rangers on motorbikes from local Indigenous communities patrol forests, listening for sounds of chain saws and using the Prey Lang smartphone app developed by Danish and Cambodian university partners to document evidence and Global Positioning System (GPS) coordinates of illegal logging.[24] While the rangers have the authority to confiscate logging equipment they encounter, they do not have the power to impose sanctions nor disrupt networks of collusion between loggers and state officials. When the compiled data from the app was presented to the Cambodian government directly to show the far-reaching extent of illicit logging, official protected areas were not expanded until the media published the data, spurring civil society groups to put pressure on the government.[25] From GFW to localized deforestation monitoring platforms, these data infrastructures rely on expertise based in the Global North and direct collaboration between foreign partners and local nonstate actors.

Environmental data infrastructures for monitoring extralegal wildlife trade—the world's fourth-largest illegal trade after humans, drugs, and counterfeited objects—are also being deployed throughout Southeast Asia. In the case of wildlife tracking, data infrastructures have both made it easier for actors to capture and trade species and emerged as a solution to that very problem. Wildlife cybercrime, in which buyers and sellers of poached flora and fauna species from across Southeast Asia use social media to conduct sales, has increased as smartphones and social media use have become more prevalent. Traders are able to use internet tools to encrypt communications, transfer funds anonymously, and deceive law enforcement by launching cyberattacks on detection software.[26] In some cases videos of wildlife slaughter uploaded by poachers themselves have led to their arrest, but an increasing amount of wildlife trafficking activity is able to switch platforms when their posts are detected.[27] Critics have pointed to the necessity of better coordination among lawmakers writing regulations at regional Association of Southeast Asian Nations scales, transnational nongovernmental organizations (NGOs), and companies that own the platforms used by traffickers.[28]

The ease with which illegal wildlife traders use social media platforms such as Facebook and Instagram to conduct business has led partnerships that work to prevent wildlife trafficking by using artificial intelligence (AI) software developed by U.S.-based researchers to scan and flag relevant social media posts. A 2018 World Bank report highlights the importance of free, open-source software, such as the Wildlife Conservation Society's Wildlife Alert App, to collect and analyze data on endangered animal species, particularly through data systems that can be accessed in situ by patrollers looking for poaching activity and report animal locations. As the report notes, these apps are most useful when they include specialized information appropriate to the places and communities they are serving, such as Vietnamese language embedded in an online platform to identify animal parts from commonly traded species.[29] In partnership with the World Conservation Society and scientists from New York's Bronx Zoo, conservationists in Vietnam have used smartphones to collect field-based data on environmental DNA in water samples, the presence of which indicates rare, hard-to-locate species such as the Hoan Kiem turtle.[30]

Overall, the complexity of AI technology for users on the ground, poor mobile phone connectivity in spots where logging and poaching are most prevalent, and the limited power of rangers to enforce regulations against the loggers reveal the limits of these new technologies to solve old resource problems.[31] The data produced by these new platforms also risks legitimizing heavy-handed law enforcement that criminalizes the rural poor without focusing on the larger drivers of environmental degradation, particularly the ways that governments and corporations are complicit in illicit activities. At the same time, the accessibility of this data to journalists and human rights activists can play an important role in holding governments accountable for environmental crimes.[32]

Histories of Surveillance and Emergent Data Infrastructure in Myanmar

Myanmar's rich and contested natural resources, its recent experiment with democracy, and the rapid rise of internet connectivity provide a timely case to examine how histories of environmental crime and exploitation shape the political ecology of data. For half a century successive military regimes fought ethnic armed organizations (EAOs), repressed the population, and extracted lucrative resources. Since the 1962 military coup successive authoritarian governments surveilled citizens with extensive networks of spies and informers and sharply curtailed freedom of speech and news from the outside world.[33] Well into the 2000s the military routinely took land from smallholder farmers and gave oil and gas revenues, mining permits, and agribusiness plantations to the wealthy and well connected. Today illicit activities such as opium production generate livelihoods—but also risks—for the rural poor while enriching elites.[34] Some government actors are active participants in illicit industries: the cross-border trades of drugs, jade, timber, and wildlife are often controlled by EAOs and the Myanmar military, producing not only environmental degradation but also overlapping and violent patterns of state and nonstate territorial control.[35]

This history of repression alongside military extraction, enrichment, and impunity made putting into place just and transparent systems of natural resource governance a key political issue in the 2010s as Myanmar undertook legal reforms, partially liberalized its economy, and elected the National League for Democracy. Curtailing illicit activities and redistributing resource wealth were central to the civilian government's promises to promote sustainable and equitable development and restore the rule of law. Increasing equity and transparency in resource governance remained a key priority until the elected government was deposed in a February 2021 military coup.

New data infrastructures for environmental monitoring that aimed to illuminate the secret, illicit, and unequal practices of resource management were enabled by expanded digital connection during Myanmar's democratic decade. Myanmar was the last Southeast Asian nation to come online, and when it did, access was severely restricted; in 1996 the military government passed a law making connecting to any computer network without authorization punishable by seven to fifteen years in prison.[36] Strict control over informational and communication technologies (ICTs) continued through government censorship and economic discrimination: in the early 2000s a SIM card could cost over US$2,000, more than half the average annual income. In 2013, when the Myanmar government opened up auctions for telecommunications licenses, the country had less than 1 percent internet penetration. Contracts were awarded to two foreign companies, Telenor and Ooredoo, in January 2014.[37] Until mid-2017 the Myanmar government controlled the country's only internet exchange point, which handles all internet data traffic in and out of the country.[38]

In 2017 the Singaporean-owned telecommunications infrastructure firm Campana signed a deal to bring the country's first privately owned undersea internet cable to Myanmar, tripling the country's available bandwidth.[39] Official government figures report that Myanmar's number of internet users rose from two million in 2014 to more than thirty-nine million in the first two years after the liberalization of the telecommunications sector, with the number of SIM cards in circulation quadrupling.[40] According to GSMA's Mobile Connectivity Index, Myanmar had 90 percent 3G coverage and 98 percent mobile penetration in 2018.[41] Somewhere between fourteen million and twenty-two million Myanmar users, 25–40 percent of the nation's population, accessed the country's most popular platform—Facebook—for news, community connection, and social mobilization. In the wake of the 2021 military coup the internet became a key site of struggle, with the military junta enforcing internet blackouts and drafting a draconian cybersecurity law to counter prodemocracy mobilization on Facebook.

The rapid rise of mobile internet usage during the democratic decade generated international excitement, with over US$2.8 billion of foreign direct investment pumped into the telecommunications sector in two years, and raised new social and political questions—for example, in critiques of Facebook's role in amplifying hate speech and extending state surveillance.[42] Expanding Myanmar's data infrastructure faces a variety of technical challenges, including the fact that during the military era Myanmar developed a separate computational encoding system (Zawgyi) incompatible with the global standard (Unicode).[43] These technical challenges become particularly potent given Myanmar's history of state surveillance, as Emad and McAuliffe note in their assessment of privacy and security in the ICT sector.[44] The December 2016 murder of a Myanmar journalist investigating illegal logging, a string of arrests targeting activists and writers for controversial Facebook posts, and extended government-imposed internet blackouts in conflict areas showed that even during civilian rule, threats remained both to environmental activism and free speech and to erecting sustainable data infrastructures.[45] Threats to both online and offline freedoms are magnified with the return of the military regime.

Mapping Platforms and Monitoring Jade

Though many of Myanmar's government offices lacked computers, digital platforms for sharing information on the environment proliferated over the democratic decade.[46] Both state and nonstate actors, often funded by international donors, expanded their knowledge production and environmental governance initiatives. These included the Myanmar Information Management Unit (MIMU), a searchable public database of development and humanitarian documents and maps created in 2007 and administered by the United Nations, and Myanmar Land, Agribusiness, and Forestry Forum (MYLAFF), a free membership-based repository of documents and presentations related to land rights established in July 2014 by a local land rights NGO.[47] Myanmar sites for Open Development Myanmar and Clean Air Myanmar, an urban air quality monitoring website, were both launched in partnership with the Yangon-based innovation lab Phandeeyar.[48] Meanwhile, One Map, a Swiss government-funded platform run through the University of Bern, worked with various government departments to centralize, standardize, and analyze spatial data. The Myanmar Forest Department began digitizing its own forest inventory in 2018 with international funding and technical expertise.[49]

These diverse initiatives for digital environmental monitoring represent a range of resources monitored, data sources and types, and intended uses. Some relied primarily on government data and served a government audience, while others sought to make satellite data or ground-sourced information more widely available. Some focused on a single resource, while others hosted a variety of maps, reports, datasets, and humanitarian announcements. Like the regional initiatives explored in the preceding section, most government and civil society platforms relied on international funding and foreign hardware, software, and analysis expertise. Foreign funds can provide critical support to monitoring initiatives, but they can also structure platforms' aims and limit their duration to donors' funding cycles. In several cases we learned about in Myanmar, foreign-conceived and funded platforms failed to catch on with local communities and were ultimately abandoned. These platforms tended to be built in English, meaning that even when a Burmese site was available, users were limited to foreigners and educated elites. International organizations often relied on government permission to work in the country, a relationship that can distort data and its availability. For example, in May 2020 a Rohingya organization accused the UN of pandering to the Myanmar government by erasing Rohingya villages from MIMU maps of areas where the Myanmar military had conducted clearance operations to expel the Muslim ethnic minority in 2017.[50]

One example of digital environmental monitoring in action is Open Jade Data (OJD), a platform launched in May 2018 by the Natural Resources Governance Institute, an international NGO, to monitor Myanmar's jade sector. The platform combines government economic data, satellite images, and photojournalism to tell the story of the world's largest jade producer. Much of Myanmar's gemstone mining and trade is extralegal, characterized by a flawed legal framework, weak institutions rife with conflicts of interest, widespread violations of license terms, and exploitative and informal labor.[51] Today the vivid green gemstone sells for more than gold in the markets of Beijing and beyond, but jade mining is dangerous work marked by atrocious conditions and widespread drug use by miners. The jade trade is not only poorly regulated but also historically connected to armed conflict in

Faxon and Goldstein

northern Myanmar, funding both the Myanmar military and ethnic militias.[52] Deforestation and environmental pollution from the mines have a direct impact on water quality and the health of surrounding populations. Fatal landslides have become increasingly common, particularly during the monsoon season. In July 2020, for example, over 168 people were killed when a wall of an open-pit mine collapsed.[53]

Estimates of the industry's worth range from US$1.5 billion to US$31 billion annually, or 50 percent of Myanmar's GDP, a range that underscores the prevalence of illicit activity and difficulties in obtaining reliable data.[54] OJD follows open-data standards to provide raw, machine-readable data sourced from government websites and Extractive Industries Transparency Initiative reports and uses open-source tools such as Datawrapper to create interactive graphics and stories that integrate data and journalistic accounts. A data story on environmental degradation due to jade mining, for example, brings together journalists' interviews, photos and videos, Google Earth satellite imagery, and government data to explore the extent and impacts of deforestation, pollution, landslides and flooding.[55] Embedded interactive images made with the Northwestern University Knight Lab's free program Juxtapose compare Google Earth images from 2013 and 2016 to show land use change, with new settlements and roads built on mine tailings that have been heaped into lakes and on forested foothills near Hpakant, Kachin State.

OJD weaves together heterogeneous data to tell the environmental, economic, and social story of a resource whose extraction has been historically characterized by violence, exploitation, and extralegality. But if the case of OJD provides a success story of moving formerly restricted data online to open-access formats, it also illuminates some common challenges of emergent data infrastructures for environmental monitoring in resource sectors marked by illegality. Three sets of challenges ultimately constrain the ability of nonstate actors to use digital tools to hold the state to account for more equitable and effective environmental governance.

First, OJD suffers from major data gaps and mismatches. For example, the quantitative data on OJD, whether sourced from the Myanmar government or international indices, is mostly limited to the legal sector, which makes up a fraction of the total trade. Platforms that rely in part on government data must contend with official records that are notoriously partial, unreliable, and restricted. Many maps remain official state secrets, and baseline data for comparisons often does not exist. The Myanmar government does not have a standardized geographic coordinate system, which means that different government ministries use widely varying spatial datasets and produce mismatched maps that do not correspond across departments or with satellite imagery. The One Map program was designed to address this problem by compiling spatial data from all departments into a single format and platform, but persuading rival ministries to agree to a standardized system proved challenging.[56] Five years after the project started the program had yet to deliver on its original promise to share government data publicly. Many of our respondents within both the government and the NGO sector noted an unwillingness to share data, a problem compounded by self-censorship of sensitive information. Ground-truthing remote or government data can be costly or impossible, particularly in areas of active armed conflict. Together these factors mean that digital platforms designed to fill data gaps continue to contend with the legal and practical problems that can perpetuate inaccurate, partial, and mismatched data.

Second, OJD struggles to process data. Data often arrives on hard copy or in scanned PDFs from government offices and must be manually entered into Excel spreadsheets. Raw data requires substantial labor to clean and make sense of internal inconsistencies, such as when three coordinates for a mining concession are clustered and the last is hundreds of miles away.[57] Users, and platforms themselves, often lack resources to analyze raw data, leading to a deluge of information without increased public understanding. A poignant example of this analysis bottleneck came in August 2018, when an irrigation dam near Myanmar's capital suddenly burst, flooding the surrounding fields and national highway. Within hours videos, photographs, and descriptions of the damage were circulating on Facebook. These digital entries were stunning evidence of the environmental destruction taking place, and yet without analysis, the amateur accounts conveyed little actionable information. One respondent working in water governance explained that finding an analyst to interpret crowdsourced flooding videos, like the ones that proliferated during the dam burst in Myanmar and another that had occurred a month earlier in neighboring Laos, had been extremely difficult.[58] The few hydrologists and engineers with relevant expertise were rarely available on demand in emergencies and often wary of making judgments on imperfect data because of fears of inaccuracies or of angering national governments for whom they often work.

As a result, international media latched onto social media accounts of disasters, while expert analysis of this crowdsourced data that could facilitate relief and recovery lagged behind. The outsourcing of analysis to foreign experts with their own interests raised concerns over data ownership and use for some environmental activists we spoke with, one of whom criticized "analyzing data in a vacuum"—the approach taken by technical actors removed from the emotional and normative stakes of resource struggles. As one foreigner working on several open data initiatives in Southeast Asia pointed out, data itself is useless without both technical analysis and culturally relevant communication. "What really, really counts is having people with GIS expertise who then pull the data for different purposes and then use it for advocacy reports and activities . . . and that's a whole investment in whole sets of expertise."[59]

Even after acquiring and analyzing the data, the ability of OJD to enact meaningful change in resource governance is far from certain, especially given entrenched inequalities in the gemstone sector. By highlighting the tremendous monetary value of legal jade and estimating the lost revenues of illegal trade, OJD's designers hope to draw attention toward the possibilities of development through more competent and ultimately equitable management of a multibillion-dollar industry.[60] Yet many platforms we studied had a limited number of users, and not all respondents we spoke with supported digital platforms. Some raised the concern that while digital data is difficult for average citizens to access and understand, foreign companies, the Myanmar military, or other bad actors were strategically positioned to access and use data to deepen legacies of surveillance and resource control. Clear chains of accountability are difficult to establish in the case of Myanmar's jade sector, where even during the years of the civilian government, mines and trade were controlled by the military, as well as by complex and shifting constellations of private companies and EAOs. In the wake of the coup jade continues to fund and fuel armed conflict across and beyond northern Myanmar.[61] This bleak turn underscores the limits of transparency and its tools. Rather than being inherently good or bad for environmental governance, data infrastructures are embedded in historical and political struggles over ethnic territory, environmental uses, and unequal harms and benefits.

Conclusions

The case of jade monitoring in Myanmar illustrates a broader trend across Southeast Asia, where new data infrastructures are emerging to track illicit resource extraction and trade and to better know and manage environments. Our analysis highlights both the possibilities and limits of digital monitoring in hybrid governance regimes with legacies of surveillance and state complicity in extralegal resource extraction. While emergent data infrastructures provide new tools for nonstate actors to counter the state's historic monopoly on information, the ability of these infrastructures to disrupt illicit economies of extraction and enrichment is far more constrained. Ultimately, the ability of nonstate actors to use digital tools to hold those in power to account rests on the responsiveness, fairness, and efficacy of governance systems, a point tragically underscored in the case of Myanmar's resurgent authoritarian regime. A political ecology of data approach highlights the continued importance of understanding situated historical relationships among citizens, the state, resources, and information in the Global South in order to assess how rapid digital connectivity influences resource access, use, and control.

Notes

1. Blok, Nakazora, and Winthereik, "Infrastructuring Environments"; Edwards, *Vast Machine*; Mol, "Environmental Governance"; Gabrys, "Smart Forests."

2. Devine and Baca, "Political Forest"; Goldstein, "Volumetric Political Forest."

3. Miller et al., "Hybrid Governance."

4. We define Southeast Asia as the region including Myanmar, Thailand, Laos, Cambodia, Vietnam, Malaysia, Singapore, Indonesia, Brunei, and the Philippines.

Faxon and Goldstein

5. See, for example, Shim's discussion in "Remote Sensing Place" of how remote sensing shaped understandings of North Korea and became a symbol of its backwardness and otherness.

6. Goldstein and Faxon, "New Data Infrastructures"; Gupta, Boas, and Oosterveer, "Transparency"; Vijge, "(Dis)Empowering Effects of Transparency."

7. Srnicek, *Platform Capitalism*.

8. Taylor, "What Is Data Justice?"

9. Cinnamon, "Attack the Data."

10. Bakker and Ritts, "Smart Earth."

11. Zuboff, *Age of Surveillance Capitalism*.

12. Xiao, "Road to Digital Unfreedom."

13. Tellman et al., "Understanding the Role," 176.

14. Tellman et al., "Understanding the Role."

15. Baker and Milne, "Dirty Money States."

16. Neimark, "Address the Roots," 139.

17. Kelly and Kelly, "Validating the Remotely Sensed Geography."

18. Following Kelly and Kelly, we use the term *extralegal* here to connote that while these activities may be illegal in contexts of international law, their illegality is less clear in certain local political and regulatory contexts.

19. INTERPOL-UNEP, *Strategic Report*.

20. Tellman et al., "Understanding the Role."

21. See Global Forest Watch, https://www.globalforestwatch.org/.

22. Barad and Jamilla, "Hutanwatch."

23. Goldstein, "Afterlives of Degraded Tropical Forests."

24. Ives, "Using Old Cellphones"; Prey Lang Community Network, "Prey Lang App."

25. Biggar, "Tackling Logging in Cambodia?"

26. Parker, "Social Media."

27. "Men Held after Monkey Killed."

28. ASEAN Post Team, "ASEAN's Illegal Wildlife Trade."

29. World Bank, "Tools and Resources."

30. Tran, "Implementation."

31. Ives, "Using Old Cellphones."

32. Neimark, "Address the Roots."

33. Fink, *Living Silence in Burma*.

34. Myanmar Opium Farmers Forum, "Statement."

35. Woods, "Rubber out of the Ashes"; Su, "Fragmented Sovereignty"; Meehan, "Fortifying or Fragmenting?"

36. Brooten, McElhone, and Venkiteswaran, "Introduction," 27.

37. Heijmans and Nyunt, "Ooredoo and Telenor."

38. Schia, "Cyber Frontier."

39. "Alcatel Joins Myanmar."

40. Nyunt, "Ministry Puts Mobile Penetration at 90 Percent."

41. GSMA, "Mobile Connectivity Index."

42. "Telecom Sector Received FDI"; Dean, "Myanmar"; Fink, *Living Silence in Burma*.

43. Hotchkiss, "Battle of the Fonts."

44. Emad and McAuliffe, "Privacy Risks."

45. A website launched by Myanmar activists to track convictions under a repressive clause of the 2013 Telecommunications Law reported 210 cases brought, most for defamation, in December 2019, with a 100 percent conviction rate. See https://www.saynoto66d.info/. See also "Myanmar: Journalist"; Beech and Hang, "Government Cut Their Internet."

46. In this section we draw on twenty-four semistructured interviews conducted by either author between January 2017 and January 2019, as well as participant observations conducted since 2014 as part of Faxon's long-term research, to briefly map diverse platforms for environmental monitoring that emerged in Myanmar between 2007 and 2019. Interviewees are cited as anonymous to preserve confidentiality.

47. MIMU, http://themimu.info/; MYLAFF, https://landportal.org/node/73478.

48. Open Development Myanmar, https://opendevelopmentmyanmar.net/; Clean Air Myanmar, "Current Air Quality Index," https://www.cleanairmm.com/.

49. Anonymous, interview by author, January 2019, Naypyitaw.

50. Kaladan Press, "ARNO."

51. Shortell, *Governing the Gemstone Sector*.

52. Courtney, "Jade."

53. Nang and Paddock, "Myanmar Jade Mine Collapse."

54. Oak, "Even with New Data."

55. Paing, "Irreparable Damage."

56. Anonymous, interview by author, January 2017, Yangon.

57. Anonymous, Skype interview by author, August 2018; anonymous, interview by author, September 2018, Yangon.

58. Anonymous, interview by author, August 2018, Yangon.

59. Anonymous, interview by author, August 2018, Yangon.

60. Anonymous, interview by author, August 2018, Yangon.

61. "Jade and Conflict."

Bibliography

"Alcatel Joins Myanmar Undersea Internet Cable Project." *Frontier Myanmar*, January 13, 2016. https://www.frontiermyanmar.net/en/alcatel-joins-myanmar-undersea-internet-cable-project/.

ASEAN Post Team. "ASEAN's Illegal Wildlife Trade Goes Online." *ASEAN Post*, March 11, 2019. https://theaseanpost.com/article/aseans-illegal-wildlife-trade-goes-online.

Baker, Jacqui, and Sarah Milne. "Dirty Money States: Illicit Economies and the State in Southeast Asia." *Critical Asian Studies* 47, no. 2 (2015): 151–76.

Bakker, Karen, and Max Ritts. "Smart Earth: A Meta-Review and Implications for Environmental Governance." *Global Environmental Change* 52 (2018): 201–11.

Barad, Richard, and Stephanie Jamilla. "Hutanwatch Spearheads Forest Data Transparency in Malaysia." *Global Forest Watch* (blog), November 26, 2019. https://blog

.globalforestwatch.org/people/hutanwatch-spearheads-forest-data-transparency
-in-malaysia.

Beech, Hannah, and Saw Hang. "The Government Cut Their Internet." *New York Times*,
July 2, 2019. https://www.nytimes.com/2019/07/02/world/asia/internet-shutdown
-myanmar-rakhine.html.

Biggar, Hugh. "Tackling Logging in Cambodia? There's an App for That." *Global Landscapes Forum* 22 (July 2019). https://news.globallandscapesforum.org/37283/tackling
-logging-in-cambodia-theres-an-app-for-that/.

Blok, Anders, Moe Nakazora, and Brit Ross Winthereik. "Infrastructuring Environments."
Science as Culture 25, no. 1 (2016): 1–22.

Brooten, Lisa, Jane McElhone, and Gayathry Venkiteswaran. "Introduction: Myanmar
Media Historically and the Challenges of Transition." In *Myanmar Media in Transition: Legacies, Challenges and Change*, edited by Gayathry Venkiteswaran, Lisa Brooten and Jane Madlyn McElhone, 1–56. Singapore: ISEAS, 2019.

Cinnamon, Jonathan. "Attack the Data: Agency, Power, and Technopolitics in South
African Data Activism." *Annals of the American Association of Geographers* 110, no.
3 (2019): 623–39.

Clean Air Myanmar. "Current Air Quality Index." Accessed June 2, 2020. https://www
.cleanairmm.com/.

Courtney, Oliver. "Jade: Myanmar's 'Big State Secret.'" *Global Witness*, October 23, 2015.
https://www.globalwitness.org/en/campaigns/oil-gas-and-mining/myanmarjade/.

Dean, Karin. "Myanmar: Surveillance and the Turn from Authoritarianism?" *Surveillance and Society* 15 no. 3/ 4 (2017): 496–505.

Devine, Jennifer, and Jenny A. Baca. "The Political Forest in the Era of Green Neoliberalism." *Antipode* 52, no. 4 (2020): 911–27.

Edwards, Paul N. *A Vast Machine: Computer Models, Climate Data, and the Politics of
Global Warming*. Cambridge MA: MIT Press, 2010.

Emad, Kamran, and Erin McAuliffe. "Privacy Risks in Myanmar's Emerging ICT Sector." In *Myanmar Media in Transition: Legacies, Challenges and Change*, edited by
Gayathry Venkiteswaran, Lisa Brooten, and Jane Madlyn McElhone, 137–50. Singapore: ISEAS, 2019.

Fink, Christina. "Dangerous Speech, Anti-Muslim Violence, and Facebook in Myanmar."
Journal of International Affairs 71, no. 1/5 (2018): 43–52.

———. *Living Silence in Burma: Surviving under Military Rule*. 2nd ed. Chiang Mai,
Thailand: Silkworm, 2009.

Gabrys, Jennifer. "Smart Forests and Data Practices: From the Internet of Trees to Planetary Governance." *Big Data & Society* 7, no. 1 (2020).

Goldstein, Jenny E. "The Afterlives of Degraded Tropical Forests: New Value for Conservation and Development." *Environment and Society: Advances in Research* 5, no.
1 (2014): 124–40.

———. "The Volumetric Political Forest: Territory, Satellite Fire Mapping, and Indonesia's Burning Peatland." *Antipode* 5, no. 1 (2020): 54–23.

Goldstein, Jenny E., and Hilary O. Faxon. "New Data Infrastructures for Environmental Monitoring in Myanmar: Is Digital Transparency Good for Governance?" *Environment and Planning E: Nature and Space* (2020).

GSMA. "Mobile Connectivity Index: Myanmar." GSM Association, accessed August 6, 2020. https://www.mobileconnectivityindex.com/#year=2018&zoneIsocode=MMR.

Gupta, Aarti, Ingrid Boas, and Peter Oosterveer. "Transparency in Global Sustainability Governance: To What Effect?" *Journal of Environmental Policy & Planning* 22, no. 1 (2020): 84–97.

Heijmans, Philip, and Aung Kyaw Nyunt. "Ooredoo and Telenor Get Licences." *Myanmar Times*, February 2, 2014. https://www.mmtimes.com/business/9433-ooredoo -telenor-get-licences.html.

Hotchkiss, Griffin. "Battle of the Fonts." *Frontier Myanmar*, March 23, 2016. https://www .frontiermyanmar.net/en/battle-of-the-fonts/.

INTERPOL-UNEP. *Strategic Report: Environment, Peace, and Security; A Convergence of Threats.* Lyon, France: INTERPOL-UN Environment Programme, 2016. https:// wedocs.unep.org/handle/20.500.11822/17008.

Ives, Mike. "Using Old Cellphones to Listen for Illegal Loggers." *New York Times*, October 15, 2019. https://www.nytimes.com/2019/10/15/climate/indonesia-logging -deforestation.html.

"Jade and Conflict: Myanmar's Vicious Circle." *Global Witness*, June 29, 2021. https://www .globalwitness.org/en/campaigns/natural-resource-governance/jade-and-conflict -myanmars-vicious-circle/.

Kaladan Press. "ARNO: UN Panders to Burma's GAD in Erasing Rohingya Villages from MIMU Maps." Burma News International, June 15, 2020. https://www.bnionline .net/en/news/arno-un-panders-burmas-gad-erasing-rohingya-villages-mimu-maps.

Kelly, Alice B., and Nina M. Kelly. "Validating the Remotely Sensed Geography of Crime: A Review of Emerging Issues." *Remote Sensing* 6, no. 12 (2014): 12723–51.

Meehan, Patrick. "Fortifying or Fragmenting the State? The Political Economy of the Opium/Heroin Trade in Shan State, Myanmar, 1988–2013." *Critical Asian Studies* 47, no. 2 (2015): 253–82.

"Men Held after Monkey Killed and Eaten in Facebook Livestream." *Guardian*, December 28, 2019. https://www.theguardian.com/world/2018/dec/28/men-arrested-vietnam -facebook-livestreaming-endangered-langur-monkey-killing.

Miller, Michelle A., Carl Middleton, Jonathan Rigg, and David Taylor. "Hybrid Governance of Transboundary Commons: Insights from Southeast Asia." *Annals of the American Association of Geographers* 110, no. 1 (2019): 297–313.

Mol, Arthur. "Environmental Governance in the Information Age: The Emergence of Informational Governance." *Environment and Planning C: Government and Policy* 24, no. 4 (2016): 497–514.

"Myanmar: Journalist Investigating Illegal Logging Killed." *Article 19*, December 14, 2016. https://www.article19.org/resources/myanmar-journalist-investigating-illegal-logging -killed/.

Myanmar Opium Farmers Forum. "Statement from the 7th Myanmar Opium Farmers' Forum: Pekhon, Southern Shan State; Declaration." Transnational Institute, May

10, 2019. https://www.tni.org/en/article/statement-from-the-7th-myanmar-opium -farmers-forum.

Nang, Saw, and Richard C. Paddock, "Myanmar Jade Mine Collapse Kills at Least 168." *New York Times*, July 2, 2020. https://www.nytimes.com/2020/07/02/world/asia /myanmar-jade-mine-collapse.html.

Neimark, Benjamin. "Address the Roots of Environmental Crime." *Science* 364, no. 6436 (2019): 139.

Nyunt, Aung Kyaw. "Ministry Puts Mobile Penetration at 90 Percent." *Myanmar Times*, July 19, 2016. https://www.mmtimes.com/business/technology/21466-ministry -puts-mobile-penetration-at-90-percent.html.

Oak, Yan N. "Even with New Data, Valuing Myanmar's Jade Industry Remains a Challenge." Open Jade Data, accessed August 7, 2020. https://openjadedata.org/Stories /how_much_jade_worth.html.

Paing, Tin Htet. "Irreparable Damage: The Environmental Impacts of Jade Mining." Open Jade Data, accessed August 8, 2020. https://openjadedata.org/Stories/jade _and_environment.html.

Parker, Stephanie. "Social Media, E-Commerce Sites Facilitate Illegal Orchid Trade." *Mongabay*, December 21, 2018. https://news.mongabay.com/2018/12/social-media -e-commerce-sites-facilitate-illegal-orchid-trade/.

Prey Lang Community Network. "Prey Lang App." Accessed August 7, 2020. https:// preylang.net/about/the-prey-lang-app/.

Schia, Niels Nagelhus. "The Cyber Frontier and Digital Pitfalls in the Global South." *Third World Quarterly* 39, no. 5 (2018): 821–37.

Shim, David. "Remote Sensing Place: Satellite Images as Visual Spatial Imaginaries." *Geoforum* 51 (2014): 152–60.

Shortell, Paul. *Governing the Gemstone Sector: Considerations for Myanmar.* New York: Natural Resource Governance Institute, 2017. https://resourcegovernance.org/sites/default /files/documents/governing-the-gemstone-sector-considerations-for-myanmar.pdf.

Srnicek, Nick. *Platform Capitalism.* Cambridge, UK: Polity, 2017.

Su, Xiaobo. "Fragmented Sovereignty and the Geopolitics of Illicit Drugs in Northern Burma." *Political Geography* 63 (2018): 20–30.

Taylor, Linnet. "What Is Data Justice? The Case for Connecting Digital Rights and Freedoms Globally." *Big Data & Society* 4, no. 2 (2017).

"Telecom Sector Received FDI of Over US$2.8 Billion." *Global New Light of Myanmar*, January 16, 2017. https://www.globalnewlightofmyanmar.com/telecom-sector -received-fdi-of-over-us2-8-billion/.

Tellman, Beth, Nicholas R. Magliocca II, B. L. Turner, and Peter H. Verburg. "Understanding the Role of Illicit Transactions in Land-Change Dynamics." *Nature Sustainability* 3, no. 3 (2020): 175–81.

Tran, Phuc. "Implementation of Smartphone-Based Environmental DNA Testing and Capture Technologies to Detect Rafetus swinhoei in the Wild." *WCS Vietnam*, news release, May 24, 2018. https://vietnam.wcs.org/News/Media-Releases/ID/11313

/Implementation-of-smartphone-based-environmental-DNA-testing-and-capture
-technologies-to-detect-Rafetus-swinhoei-in-the-wild.aspx.

Vijge, Marjanneke J. "The (Dis)Empowering Effects of Transparency beyond Information
Disclosure: The Extractive Industries Transparency Initiative in Myanmar." *Global
Environmental Politics* 18, no. 1 (2018): 13–32.

Woods, Kevin. "Rubber out of the Ashes: Locating Chinese Agribusiness Investments
in 'Armed Sovereignties' in the Myanmar–China Borderlands." *Territory, Politics,
Governance* 7, no. 1 (2018): 79–95.

World Bank. *Tools and Resources to Combat Global Wildlife Trade*. Washington DC:
World Bank, 2018. http://pubdocs.worldbank.org/en/389851519769693304/24691
-Wildlife-Law-Enforcement-002.pdf.

Xiao, Qiang. "The Road to Digital Unfreedom: President Xi's Surveillance State." *Journal of Democracy* 30, no. 1 (2019): 53–67.

Zuboff, Shoshana. *The Age of Surveillance Capitalism: The Fight for a Human Future at
the New Frontier of Power*. New York: PublicAffairs, 2019.

THIRTEEN

Data Gaps

PENGUIN SCIENCE AND PETROSTATE FORMATION
IN THE FALKLAND ISLANDS (MALVINAS)

James J. A. Blair

When it comes to petroleum, we tend to associate seabirds with shocking images of oil-soaked feathers, contaminated in the wake of a toxic spill or disastrous blowout. However, seabirds and the data they generate have also become key actors in the crafting of environmental governance systems for new oil-drilling projects. This chapter examines how penguin science has come to play a significant role in the development of data infrastructure for an offshore oil frontier.[1] It does so through ethnographic description of scientific initiatives to fill "data gaps."[2] This is a term that scientists use to label unexplored areas of research—in this case, as they pertain to biodiverse species in an underanalyzed geographic location: ocean waters surrounding the Falkland Islands (Malvinas in Spanish).[3]

Positioned near the tip of South America, this remote archipelago has served historically as a gateway to Antarctica. A violent 1982 war between Argentina and the United Kingdom cemented British control over the Falklands, but disagreements over commercial fishing and recent discoveries of oil by British companies have sparked a renewed sovereignty dispute. The island chain is currently a British overseas territory, but the Argentine government continues to claim that it is an integral part of its national territory, issuing legal threats to oil companies operating in the South Atlantic.

This chapter traces the emergence of new data infrastructure to support oil exploration in this geopolitical hotspot. Drawing on twenty months of ethnographic fieldwork and historical research in the Falklands, Argentina, and the United Kingdom, I describe how marine ecologists, funded by both the British Falkland Islands government (FIG) and its licensed oil companies, tag penguins with tracking sensors. Data representing the penguins' foraging patterns are intended to fill gaps in knowledge, giving oil companies social license to operate and ideally informing environmental management and monitoring. Even though oil production poses inherent threats to seabird survival, I argue that new data infrastructure, built on information about penguins and other marine fauna, has boosted oil development in a popular framing of good environmental governance.

While much ink has been spilled on the battle history and memory of the 1982 military conflict between the United Kingdom and Argentina, little has been written about the more recent formation of a petrostate in the Falklands.[4] Fernando Coronil has shown how the modern state acts as a landowner in other oil-rich territories, such as Venezuela, generating money from nature by renting access to the nation's subsoil.[5] Moreover, oil is not simply a "cursed" resource that leads inevitably to authoritarian rule, for as Timothy Mitchell has argued, carbon energy has also served as a means of making democratic claims since the era of decolonization.[6] Contested claims to natural resources have now made the Falklands a critical case for examining the significance of environmental data governance systems in democratic societies that consent to remain under their former colonial administrator's flag. Since the war, enhanced British access to and control over marine resources have transformed the South Atlantic into a lucrative commercial fishing hub and now an offshore oil frontier.

In 2010 Rockhopper Exploration, one of the Falklands' small British oil licensees, named after a local penguin species, discovered commercial amounts of oil (the equivalent of four hundred million barrels) in the Sea Lion well of the North Falkland Basin. Rockhopper's partner Premier Oil, a larger and more experienced British independent firm, prepared to exploit it.[7] This was highly controversial in Argentina, where the rise of Presidents Néstor Kirchner and Cristina Fernández de Kirchner ushered in a left-leaning populist government, which sought to consolidate power around the national cause of regaining sovereignty over the South Atlantic and its resources.[8] In response Falkland Islanders held a 2013 referendum on self-determination, voting 99.8 percent in favor of remaining British, with just 3 "no" votes out of 1,517 valid ballots.[9] By claiming self-determination, the islanders were building an oil protectorate through popular consent to British sovereignty. British imperial nostalgia and right-wing populism, which the Falklands referendum conjured, also motivated UK citizens to vote to leave the European Union in 2016. However, Brexit has ironically imposed costly tariffs on Falklands fishing exports and cut substantial funding for biodiversity research, thus raising the stakes for a future oil industry linked to environmental science in the islands.

To shed light on an underresearched aspect of this emerging extractive economy, this chapter analyzes how the dual prospects of oil and sovereignty influence scientists collecting data on the interactions between penguin feeding and exploration drilling.[10] The affordability and availability of wireless tracking sensors with enhanced storage capacity have made data a "key mediator" in environmental governance.[11] Yet social scientific analyses of animal tracking, particularly that involving birds and marine animals, have highlighted how remote sensing may serve diverse political-ecological interests.[12] A significant divide exists between the values of wildlife conservation on the one hand and those of ecological science on the other. While some actors seek to use tracking data to spread awareness through advocacy, others may be motivated by public or private interests in surveillance data infrastructures. Before discussing further the relationship between oil interests and remote-sensing technology, I first offer a brief vignette about a wildlife conservationist who was determined to collect baseline data on foot or with aerial photography to advocate for protecting penguins from extractive industries in the Falklands.

March of the Penguin Scientist: Contending with Populist Resource Regimes

In his self-published exposé, *The Falklands Regime*, independent British penguin biologist Mike Bingham tells the impassioned story of his fight to save the Falklands' penguin population from decline. *Regime* is an ironic allusion to the kind of antidemocratic dictatorships that the 1982 Falklands/Malvinas War was ostensibly supposed to have ousted when British prime minister Margaret Thatcher ordered her military to recover the islands from the Argentine junta that had occupied the archipelago for seventy-four days. The British victory in the

South Atlantic established Thatcher as the formative figure responsible for the persistent ideology that Stuart Hall calls "authoritarian populism," a conjuncture of state domination with popular consent.[13] Thatcher used the spectacle of human sacrifice in the war to leverage the United Kingdom's patriotic political mood to fight so-called enemies within, including organized miners' unions and criminalized Black Britons.[14] The sensational cover photo of Bingham's book displays a beach covered in corpses that may seem reminiscent of the widely circulated images of carnage left on the 1982 battlefields. However, on Bingham's cover, instead of human soldiers, the dead bodies were a gruesome array of penguin carcasses that had washed ashore on the Falklands in the 1990s.[15]

Bingham places the blame for this apparent penguin decimation squarely on the emergence of a prosperous commercial fishing industry, which he claims had removed the seabirds' main food source.[16] Moreover, Bingham's early resistance not just to commercial fishing but also to offshore oil drilling, remains one of the only public acts of defiance against local conservation groups, which have had entrenched interests in the Falklands' booming extractive industries. His whistleblowing earned Bingham personal threats from political and business elites on the islands. He accused the government of a litany of coercive acts, including attempted bribery; false accusations of data theft, burglary, and fraud; framing with possession of contraband pornographic film; planting firearms; delivering death threats and malicious phone calls; committing perjury; fabricating documents; and sabotaging a vehicle. Bingham ultimately went into exile in Argentina and later Chile, but only after taking his freedom-of-speech case all the way to the British Supreme Court in 2003—and winning.[17]

Bingham's controversial case throws light on entanglements between penguin science and authoritarian populism that I explore here by analyzing the role of petrostate formation in the development of new environmental data infrastructures in the Falklands. This chapter builds on a growing body of literature on the connection between authoritarian populism and environmental governance by employing scholarship at the interface of political ecology of data as well as science, technology, and society (STS) to better understand the actions of experts in an emergent oil frontier.[18] Across oil frontiers, energy companies attempt to make permanent infrastructure invisible, as exemplified in underground pipelines.[19] Aboveground oil infrastructures, such as "man camps" and offshore oil rigs, make petroleum's presence seem short-lived, obscuring lasting forms of dispossession and toxic contamination.[20] Here I contend that by performing scientific rigor and transparency, *data* infrastructures also authorize pollution of the marine environment in oil frontiers.

The current political-ecological dynamics of data infrastructure in the Falklands extend from a wider history of imperial science in the South Atlantic and settler colonialism in the Southern Cone.[21] A late colonial frontier, the Falklands/Malvinas archipelago served initially as a staging ground for the conquest and settlement of Patagonia.[22] A dearth of scientific data on the South Atlantic has also made the islands a "frontier" for knowledge production. During the so-called Heroic Age of Antarctic Exploration, scientists used the Falklands primarily as a stepping-stone to support voyages farther south, notably Sir Ernest Shackleton's Imperial Trans-Antarctic Expedition on the *Endurance*. At the height of the British Empire, the Falklands served as an administrative center for regional science, with a series of research stations being established for the Falkland Islands Dependencies Survey, now the British Antarctic Survey.[23] Nonetheless, the South Pole's perceived importance for imperial science left the South Atlantic underresearched by comparison.

Since the 1982 war and subsequent marine resource commodity boom, two competing state projects have sought to compound science with populism to promote different data imaginaries and fill the apparent void of research in the South Atlantic.[24] In 2014 Argentina's Ministry of Science, Technology and Productive Innovation launched its Pampa Azul (Blue Pampa) campaign. Pampa Azul was an initiative of the presidency of Cristina Fernández de Kirchner, who understood the popular Malvinas sovereignty cause as a fundamental authoritarian-democratic value of Peronism—arguably the original ideology of populism.[25] Seeking to renovate Argentina's popular national origin story, Kirchner's government attempted to rebrand the Pampa, a place name that typically refers to the agricultural interior of Argentina, incorporated into the nation through the genocidal Desert Campaigns to eliminate Indigenous peoples from the territory in the 1830s and 1880s. Instead, the Argentine government began to suggest that the Atlantic Ocean was a long-neglected Pampa Azul, rich in biodiverse species and natural resources available for scientific research and commercial development. Moreover, because the Pampa Azul campaign is primarily framed as a public research program, the government has been able to wrap its geopolitical machinations for the South Atlantic in a cloak of science. Pampa Azul's slogan is "Scientific knowledge in service of national sovereignty."

A competing research center formed within the Falklands, which promoted a contending "sociotechnical imaginary" of the South Atlantic.[26] With initial backing from the FIG, the mission of the South Atlantic Environmental Research Institute (SAERI) is to grow a "knowledge economy" in the Falklands. The islands had long served as a staging post for research in Antarctica, yet little was known about the ecological dynamics of many species in the South Atlantic, so SAERI aimed to contribute new high-impact research on marine ecology, oceanography, and climate geology. SAERI is now registered as a charitable incorporated organization and is a 100 percent shareholder in the SAERI (Falklands) Limited trading subsidiary. Current and former government directors continue to serve as advisors and trustees, but oil consultancy contracts have become a key source of private capital for expansion of the institute's scope of environmental governance. SAERI's growing data infrastructure has come to serve as a hub for scientific research throughout the British overseas territories in the South Atlantic, and even in the Caribbean and Southern Africa. The institute has also been instrumental in FIG diplomacy to promote the islanders' self-determination claim, thus directly supporting petrostate formation based on popular consent to British authority.

SAERI rented me office space throughout my fieldwork in the Falklands, which gave me an inside perspective on its activities, procedures, and operations. During my research in the South Atlantic, SAERI included me in its events and welcomed my critical and applied perspective. I participated in meetings of the Falkland Islands Offshore Hydrocarbons Environmental Forum, an advisory group.[27] I also attended public meetings and provided comments on a new marine spatial plan for the South Atlantic. Toward the end of my fieldwork, SAERI hosted a public talk that I gave for local residents. And after I had departed the islands, SAERI invited me to participate in a workshop on data analysis and marine spatial planning at the University of Cambridge. FIG also invited me to apply to lead a socioecological impact monitoring study, but since the oil licensees were partial funders—who were considered to be operating illegally according to Argentine law—this may have risked fines and imprisonment of up to fifteen years in Argentina, so I politely declined. However, as a result of the workshop in Cambridge, I served as an external reviewer and provided extensive comments for the Phase I Report of SAERI's Gap Project. SAERI welcomed my suggestions and has made efforts to address them. Nevertheless, scientific studies of penguins and other aspects of marine ecology have generally propelled rather than deterred petrostate formation in the South Atlantic.

The Gap Project: Filling the Oil Frontier with Data

With joint funding from the FIG and the Falkland Islands Petroleum Licensees Association, SAERI launched the Gap Project to complete its environmental review by "first oil," an indeterminate starting point of petroleum production.[28] Here I focus primarily on the Gap Project's penguin-tracking work. This research was designed to go beyond the baseline surveys that Bingham had completed earlier as a conservation officer with Falklands Conservation, an affiliate of BirdLife International. Wendi, one of the Gap Project's lead scientists, created a new penguin-tracking program to understand how their feeding synchronizes with the market for oil.[29] For each penguin colony, Wendi was interested in patterns of foraging, such as differences in age and sex, as well as consistencies in the annual cycle that might provide a more comprehensive understanding of the penguins' interaction with oil work.

To track the penguins, Wendi primarily used global location sensing (GLS) tags, which were attached to the birds' ankles to record light levels twice per day, and Global Positioning System (GPS) satellite tracker tags, glued onto their backs. The latter are far more precise, but the former are relatively inexpensive, so most of the tags used

20. Penguin colony at Cape Bougainville. Courtesy of author.

were G L S. There are certain efficiencies that make the G L S tags less expensive than G P S trackers. First, their spatial resolution is poor, so they are only accurate beyond one hundred kilometers. If penguins travel within that distance from their colony, the tags are not particularly useful because of the large margin of error. The recorded light levels allow researchers to work out latitude and longitude using an algorithm, but multiple sunlight readings per day cause aberrations. Extensive data cleaning during analysis makes statistics difficult to run. Second, these tags archive data, which means that they are not detectable remotely and need to be physically retrieved for analysis.

Together with Wendi and two other researchers, I volunteered to remove location sensors that had been attached to rockhopper penguins at Cape Bougainville on East Falkland Island, a scenic coastal area (fig. 20). Wendi selected this site, among others, because she expected them to be most affected by a potential uncontrolled oil spill at Premier's Sea Lion well. We spent most of the day staring at the rockhoppers' pink ankles, searching for any sign of the black band and green disk of the G L S tag. We took turns prodding gently at the rockhoppers' ankles with poles, and if we found a sensor present, we isolated the tagged bird with a fishing net. We then brought the birds to

21. Penguin-tagging process. Courtesy of Amélie Augé.

Wendi, who held them in her lap for me to cut off the tags from their ankles (fig. 21). For each bird, I recorded the number-letter label on the device and the apparent sex. I then plucked four feathers as samples for isotope tests, which would be used to analyze diet and other biological features. Next, I ran a small blue bungee cord under the bird's belly and attached it to a hanging scale to measure the weight. Finally, Wendi reintroduced the bird to the rest of the colony.

We carried on with this method with steady success, locating and processing nine out of the twenty previously tagged birds. Data from the Gap Project would ostensibly inform new strategies to minimize and mitigate impact in environmental management plans, but Wendi acknowledged, "In all honesty, I don't think there's any way they'll stop going ahead with the drilling."

With its pioneering role in the Gap Project, SAERI was well positioned to contribute new local knowledge to FIG's vetting process for the oil-drilling projects. Unlike other cases in which third-party consultants produced cursory or misleading environmental impact assessments in mining and energy projects, SAERI's researchers sought to contribute rigorous science and go beyond simply meeting accepted standards, through data infrastructure and proactive programs like the Gap Project, with considerable support from the FIG and industry. However, as the institute began to distance itself from government and became incorporated as a separate charitable organization and trading subsidiary, SAERI not only reviewed environmental impact statements as the regulator but also accepted contracts from the oil companies to produce them. This shift from the use of knowledge for environmental management to industry-affiliated research signals what Jenny Goldstein calls "divergent expertise."[30] Conflicts of interest may be inevitable in such a small, remote island community. In this case SAERI found itself in an awkward position because it remained partially government-funded. Oil company managers who spoke with me focused on what they were trying to achieve and asserted that SAERI likely provided the best possible information and knowledge on the subject.

Nonetheless, important gaps remained that the penguin-tracking project has not been able to address. Perhaps most alarming is that the contested regional politics of the sovereignty dispute and Argentina's contradictory claims over maritime zones are not discussed. The penguin-tracking data indicates that multiple species, including rockhoppers, migrate directly through the Sea Lion drilling area and then far into Argentine (and even Uruguayan) waters.[31] From the perspective of the FIG and its oil licensees, the baseline population and overall proportion of penguins are both properly monitored. However, according to the Argentine government, this may merit enforcement of domestic laws against drilling. This highly controversial aspect of penguin migration was not factored into SAERI's assessment of drilling's transboundary impact in its environmental impact statements. Given the contending populist sovereignty campaigns, the FIG has acknowledged that a spill passing into Argentina's zone would be a *political* disaster. But Argentina claims the entire continental shelf (the Pampa Azul) as its national maritime area. Even though they assert popular consent to British authority through the islanders' self-determination claim, FIG legislators have not sought transboundary consent with Argentina because they view it as unachievable when the actual boundaries are not even agreed upon.

Given that their drilling activities are considered illegal according to Argentine domestic law, oil-drilling operators are not likely to consult these neighbors. They are active members in a global network called Oil Spill Response Limited, which has an office in Brazil, but it could take weeks before support vessels arrive to control an accidental spill and rescue wildlife. In this way, penguin data has become mutually constitutive with the Falklands' emerging petrostate. An authoritarian populist government—shaped by the desire for natural resources—is itself furthering oil interests through environmental science, while leaving the seabirds' cross-border habitat vulnerable to potential contamination.[32]

Conclusions

With the establishment of SAERI, the FIG has sought to anchor scientific knowledge production of the South Atlantic in the Falklands. Ethnographic analysis of SAERI's Gap Project reveals how data generated from penguin sensors may be understood as an actor that authorizes resource managers to practice continuity rather than change in environmental governance. This chapter suggests that data infrastructure may run counter to restrictive or prohibitive approaches to environmental governance, when we attend to how extractive capitalism can be rooted in authoritarian populism. In spite of considerable technological innovation and data collection, the course of resource exploitation remains undeterred, without environmental concern, political commitment, or geopolitical cooperation. This chapter has shown how data infrastructure may perform classification and compliance while boosting rather than challenging petrostate formation.

In the case of the Gap Project, penguin science in support of extractive industries suggests that remote-sensing technology may speed up rather than slow down development. Filling gaps with geospatial data about penguin migration patterns does not alter the trajectory of oil exploration—it just legitimizes it. This was an observation that some scientists I interviewed confirmed, even if they deemed the offshore oil industry unnecessary in a remote island chain with a small population of residents enjoying a relatively high standard of living. As they discover new resources, oil firms and petrostates are continually making new frontiers; as some gaps appear to close, new ones open. There remains a discrepancy between what data gap analyses reveal and what is important for sustainable governance and environmental justice.

Despite the sovereignty dispute, it is noteworthy that both the United Kingdom and Argentina have declared a climate emergency, and there is an increasing probability that offshore oil projects will become stranded assets. Nonetheless, at the time of my research, the global price of oil and the efficient availability of extractable materials in British-controlled territories still outweighed risk avoidance. This case thus offers a parable for human and nonhuman complicity in geopolitical conflict, regional pollution, and ultimately our current climate and ecological crises. Tracked for the purposes of science, statecraft, and industry, penguins and their data have become key actors in environmental governance, yet in the Falklands/Malvinas they serve the interests of an emerging petrostate. Given the potential hazards of oil production on marine ecosystems, these specimens have become conscripts in their own potential extirpation.

Notes

1. Watts, "Oil Frontiers." On the general ideology of the "frontier," see Cronon, "Trouble with Wilderness"; Grandin, *End of the Myth*.

2. Blair, "Tracking Penguins, Sensing Petroleum."

3. The sovereignty dispute has politicized the English or Spanish name for the islands. For ease of reading, I attempt to remain true to ethnographic or historical context for place names.

4. Freedman and Gamba-Stonehouse, *Signals of War*; Lorenz, *Las Guerras Por Malvinas*; Guber, *De Chicos a Veteranos*; Benwell, "Encountering Geopolitical Pasts."

5. Coronil, *Magical State*.

6. Weszkalnys, "Cursed Resources"; Mitchell, *Carbon Democracy*.

7. Premier Oil suspended production at Sea Lion amid the global glut of oil supply in 2020. Premier then became part of Harbour Energy and merged with Chrysaor Holdings. However, Harbour backed out of the Sea Lion project when the Argentine government announced sanctions in 2021.

8. Finchelstein, *From Fascism to Populism*.

9. According to a 2016 census 80 percent of 3,200 total residents consider their national identity to be Falkland Islander or British. There is an influx of migrants from St. Helena (8 percent) and Chile (5 percent). The Mount Pleasant Royal Air Force base includes 359 people.

10. Dodds and Benwell, "More Unfinished Business."

11. Ascui, Haward, and Lovell, "Salmon, Sensors, and Translation."

12. See Mitman, "When Nature Is the Zoo"; Haraway, *When Species Meet*; Benson, *Wired Wilderness*; Ray, "Rub Trees, Crittercams"; Whitney, "Domesticating Nature?"; Stokland, "Field Studies in Absentia"; Gabrys, *Program Earth*; Gabrys, "Sensors and Sensing Practices."

13. Hall, "Popular-Democratic vs. Authoritarian Populism"; Hall, "Authoritarian Populism"; Hall, "Empire Strikes Back 1982."

14. Hobsbawm, "Falklands Fallout"; Gilroy, *"There Ain't No Black,"* 51–55; Milne, *Enemy Within*; Hall, "Empire Strikes Back 1982."

15. Bingham excelled at self-publicizing by appealing to popular affection for penguins. Yet *The Falklands Regime* has a conspiratorial tone, and Bingham's numerous critics point to inaccuracies, simplifications, and exaggerations.

16. Bingham, "Distribution, Abundance and Population Trends"; Bingham, "Decline of Falkland Islands Penguins." FIG Fisheries has also adopted mitigation measures to reduce bird mortality, in accordance with the Agreement on the Conservation of Albatrosses and Petrels. Bingham's critics suggest that while there may be overlap between penguin foraging and commercial fishing, global changes in ocean temperature were the principal cause of penguin population declines. See Croxall, McInness, and Prince, "Status and Conservation of Seabirds"; Clausen and Pütz, "Recent Trends"; Pütz et al., "Reevaluation"; Dehnhard et al., "Survival of Rockhopper Penguins."

17. Bingham left his Falklands research behind after his triumphant court case. He had lived there for eleven years, from 1993 to 2004. Currently, he focuses his monitoring activities primarily in the Chilean Island of Magdalena, as well as Cabo Vírgenes, Argentina, both no-fishing zones. Mike Bingham, personal communication with author, 2014.

18. Goldman, Nadasdy, and Turner, *Knowing Nature*; McCarthy, "Authoritarianism, Populism."

19. Barry, *Material Politics*.

20. Limbert, *In the Time of Oil*; Appel, "Offshore Work"; Weszkalnys, "Geology, Potentiality, Speculation."

21. Blair, "Settler Indigeneity."

22. In 1829 Buenos Aires named Luis Vernet, a merchant originally from Hamburg, the political and military commander of the Malvinas. Vernet inherited rights over feral cattle that flourished after the abandonment of French (1764–67), British (1765–70 and 1771–74), and Spanish (1767–1811) temporary settlements. However, in response to Vernet's detainment of three U.S. sealing schooners, the American military destroyed his settlement, clearing the way for the British to reclaim sovereignty on January 2, 1833.

23. Dodds, *Pink Ice*.

24. Blair, "South Atlantic Universals."

25. Finchelstein, *From Fascism to Populism*; James, *Resistance and Integration*; Laclau, *Politics and Ideology*.

26. Jasanoff and Kim, *Dreamscapes of Modernity*; Blair, "South Atlantic Universals."

27. This forum does not have any specific powers, but it holds biannual meetings for stakeholders, including oil companies, civil society, and government officials.

28. Weszkalnys, "Geology, Potentiality, Speculation."

29. A pseudonym has been used to preserve anonymity.

30. Goldstein, "Knowing the Subterranean."

31. See Baylis et al., "Important At-Sea Areas," fig. 2.

32. Alvarez León and Gleason, "Production, Property."

Bibliography

Alvarez León, Luis F., and Colin J. Gleason. "Production, Property, and the Construction of Remotely Sensed Data." *Annals of the American Association of Geographers* 107, no. 5 (September 2017): 1075–89.

Appel, Hannah. "Offshore Work: Oil, Modularity, and the How of Capitalism in Equatorial Guinea." *American Ethnologist* 39, no. 4 (2012): 692–709.

Ascui, Francisco, Marcus Haward, and Heather Lovell. "Salmon, Sensors, and Translation: The Agency of Big Data in Environmental Governance." *Environment and Planning D: Society and Space* 36, no. 5 (October 2018): 905–25.

Barry, Andrew. *Material Politics: Disputes along the Pipeline.* Oxford: Wiley-Blackwell, 2013.

Baylis, Alastair M. M., Megan Tierney, Rachael A. Orben, Victoria Warwick-Evans, Ewan Wakefield, W. James Grecian, Phil Trathan et al. "Important At-Sea Areas of Colonial Breeding Marine Predators on the Southern Patagonian Shelf." *Scientific Reports* 9, no. 1 (June 2019): 8517.

Benson, Etienne. *Wired Wilderness: Technologies of Tracking and the Making of Modern Wildlife.* Baltimore: Johns Hopkins University Press, 2010.

Benwell, Matthew C. "Encountering Geopolitical Pasts in the Present: Young People's Everyday Engagements with Memory in the Falkland Islands." *Transactions of the Institute of British Geographers* 41, no. 2 (April 1, 2016): 121–33.

Bingham, Mike. "The Decline of Falkland Islands Penguins in the Presence of a Commercial Fishing Industry/La Disminución de Los Pinguinos de Las Islas Falklands En La Presencia de Actividades de Pesca Comercial." *Revista Chilena de Historia Natural* 75 (2002): 805–18.

———. "The Distribution, Abundance and Population Trends of Gentoo, Rockhopper and King Penguins in the Falkland Islands." *Oryx* 32, no. 3 (July 1, 1998): 223–32.

———. *The Falklands Regime.* Bloomington IN: Environmental Research Unit, 2005.

Blair, James J. A. "Settler Indigeneity and the Eradication of the Non-Native: Self-Determination and Biosecurity in the Falkland Islands (Malvinas)." *Journal of the Royal Anthropological Institute* 23, no. 3 (2017): 580–602.

———. "South Atlantic Universals: Science, Sovereignty and Self-Determination in the Falkland Islands (Malvinas)." *Tapuya: Latin American Science, Technology and Society* 2, no. 1 (September 2019): 220–36.

———. "Tracking Penguins, Sensing Petroleum: 'Data Gaps' and the Politics of Marine Ecology in the South Atlantic." *Environment and Planning E: Nature and Space* (October 2019).

Clausen, Andrea Patricia, and Klemens Pütz. "Recent Trends in Diet Composition and Productivity of Gentoo, Magellanic and Rockhopper Penguins in the Falkland

Islands." *Aquatic Conservation: Marine and Freshwater Ecosystems* 12, no. 1 (January 2002): 51–61.

Coronil, Fernando. *The Magical State: Nature, Money, and Modernity in Venezuela.* Chicago: University of Chicago Press, 1997.

Cronon, William. "The Trouble with Wilderness; or, Getting Back to the Wrong Nature." *Environmental History* 1, no. 1 (1996): 7–28.

Croxall, J. P., S. J. McInness, and P. A. Prince. "The Status and Conservation of Seabirds at the Falkland Islands." In *Status and Conservation of the World's Seabirds*, edited by J. P. Croxall, P. G. H. Evans, and R. W. Schreiber, 271–93. Cambridge: International Council for Bird Preservation, 1984.

Dehnhard, Nina, Maud Poisbleau, Laurent Demongin, Katrin Ludynia, Miguel Lecoq, Juan F. Masello, and Petra Quillfeldt. "Survival of Rockhopper Penguins in Times of Global Climate Change." *Aquatic Conservation: Marine and Freshwater Ecosystems* 23, no. 5 (October 2013): 777–89.

Dodds, Klaus. *Pink Ice: Britain and the South Atlantic Empire.* London: I. B. Tauris, 2002.

Dodds, Klaus, and Matthew C. Benwell. "More Unfinished Business: The Falklands/Malvinas, Maritime Claims, and the Spectre of Oil in the South Atlantic." *Environment and Planning D: Society and Space* 28 (2010): 571–80.

Finchelstein, Federico. *From Fascism to Populism in History.* Oakland: University of California Press, 2017.

Freedman, Lawrence, and Virginia Gamba-Stonehouse. *Signals of War: The Falklands Conflicts of 1982.* London: Faber & Faber, 1990.

Gabrys, Jennifer. *Program Earth: Environmental Sensing Technology and the Making of a Computational Planet.* Minneapolis: University of Minnesota Press, 2016.

———. "Sensors and Sensing Practices: Reworking Experience across Entities, Environments, and Technologies." *Science, Technology, & Human Values* 44, no. 5 (July 2019): 723–36.

Gilroy, Paul. *"There Ain't No Black in the Union Jack": The Cultural Politics of Race and Nation.* Chicago: University of Chicago Press, 1987.

Goldman, Mara, Paul Nadasdy, and Matt Turner, eds. *Knowing Nature: Conversations at the Intersection of Political Ecology and Science Studies.* Chicago: University of Chicago Press, 2011.

Goldstein, Jenny E. "Knowing the Subterranean: Land Grabbing, Oil Palm, and Divergent Expertise in Indonesia's Peat Soil." *Environment and Planning A* 48, no. 4 (April 2016): 754–70.

Grandin, Greg. *The End of the Myth: From the Frontier to the Border Wall in the Mind of America.* New York: Metropolitan Books, 2019.

Guber, Rosana. *De Chicos a Veteranos: Nación y Memorias de La Guerra de Malvinas.* Buenos Aires: Colección La Otra Ventana, 2009.

Hall, Stuart. "Authoritarian Populism: A Reply." *New Left Review*, no. 151 (May–June 1985): 115–24.

———. "The Empire Strikes Back 1982." In *Selected Political Writings: The Great Moving*

Right Show and Other Essays, edited by Sally Davison, David Featherstone, Michael Rustin, and Bill Schwarz, 200–206. Durham NC: Duke University Press, 2017.

———. "Popular-Democratic vs. Authoritarian Populism: Two Ways of 'Taking Democracy Seriously.'" In *Marxism and Democracy*, edited by Alan Hunt, 157–85. London: Lawrence and Wishart, 1980.

Haraway, Donna. *When Species Meet*. Minneapolis: University of Minnesota Press, 2007.

Hobsbawm, Eric. "Falklands Fallout." *Marxism Today* 26, no. 1 (January 1983): 13–19.

James, Daniel. *Resistance and Integration: Peronism and the Argentine Working Class, 1946–1976*. Cambridge: Cambridge University Press, 1988.

Jasanoff, Sheila, and Sang-Hyun Kim, eds. *Dreamscapes of Modernity: Sociotechnical Imaginaries and the Fabrication of Power*. Chicago: University of Chicago, 2015.

Laclau, Ernesto. *Politics and Ideology in Marxist Theory: Capitalism, Fascism, Populism*. London: New Left Books, 1977.

Limbert, Mandana. *In the Time of Oil: Piety, Memory, and Social Life in an Omani Town*. Stanford CA: Stanford University Press, 2010.

Lorenz, Federico. *Las Guerras por Malvinas*. Buenos Aires: Edhasa, 2006.

McCarthy, James. "Authoritarianism, Populism, and the Environment: Comparative Experiences, Insights, and Perspectives." *Annals of the American Association of Geographers* 109, no. 2 (March 2019): 301–13.

Milne, Seumas. *The Enemy Within: The Secret War against the Minors*. 4th ed. London: Verso, 2014.

Mitchell, Timothy. *Carbon Democracy: Political Power in the Age of Oil*. London: Verso, 2011.

Mitman, Gregg. "When Nature Is the Zoo: Vision and Power in the Art and Science of Natural History." *Osiris* 11 (1996): 117–43.

Pütz, Klemens, Andrea P. Clausen, Nic Huin, and John P. Croxall. "Re-evaluation of Historical Rockhopper Penguin Population Data in the Falkland Islands." *Waterbirds* 26, no. 2 (June 2003): 169–75.

Ray, Sarah Jaquette. "Rub Trees, Crittercams, and GIS: The Wired Wilderness of Leanne Allison and Jeremy Mendes' Bear 71." *Green Letters* 18, no. 3 (September 2014): 236–53.

Stokland, Hakon B. "Field Studies in Absentia: Counting and Monitoring from a Distance as Technologies of Government in Norwegian Wolf Management (1960s–2010s)." *Journal of the History of Biology* 48 (2015): 1–36.

Watts, Michael. "Oil Frontiers: The Niger Delta and the Gulf of Mexico." In *Oil Culture*, edited by Ross Barrett and Daniel Worden, 189–210. Minneapolis: University of Minnesota Press, 2014.

Weszkalnys, Gisa. "Cursed Resources, or Articulations of Economic Theory in the Gulf of Guinea." *Economy and Society* 40, no. 3 (2011): 345–72.

———. "Geology, Potentiality, Speculation: On the Indeterminacy of First Oil." *Cultural Anthropology* 30, no. 4 (November 2015): 611–39.

Whitney, Kristoffer. "Domesticating Nature? Surveillance and Conservation of Migratory Shorebirds in the 'Atlantic Flyway.'" *Studies in History and Philosophy of Science Part C: Studies in History and Philosophy of Biological and Biomedical Sciences* 45 (March 2014): 78–87.

FOURTEEN

Data Structures, Indigenous Ontologies, and Hydropower in the U.S. Northwest

Corrine Armistead

D ata structures—frameworks or sets of rules dictating how individual pieces of information are stored and linked together in digital databases—delimit what can and cannot be known through data.[1] Existing digital databases, software, and methods, specifically those that encode nature and shape environmental analysis, evolved not in a vacuum, but within larger political-economic frameworks and embedded cultural values.[2] These tools remain tied to the tenets of resource extraction, exclusion, and accumulation that spurred their development. As a result, the use of currently constructed software in environmental decision-making forecloses opportunities to expand how we see landscapes through digital data and instead continues to privilege dominant worldviews.[3]

This chapter explores these limitations through analysis of one database depicting energy generation potential in the United States. By mapping the Columbia River Basin's hydropower potential in the U.S. Pacific Northwest, it engages with data structure as one means by which nature, as a resource, is defined, managed, and extracted. It argues for using a graph data structure composed of nodes and relationships—as opposed to the more conventional but rigid relational data structure—to reshape measurement and analysis through the processes of storing and querying social and ecological data inputs. Using a graph database shifts data storage and analysis from assumptions of unique and discrete data to a framing of connection and relationship among data. Foregrounding relationships is a concept central to many Indigenous ontologies and a way to imagine how water and energy flow through our spaces, industries, and peoples.[4]

The following section describes the two data structures considered in this case study: relational and graph. Then hydropower potential databases are situated in the larger context of environmental decision-making and capitalist accumulation. Using the example of data assembled to map hydropower potential and associated environmental concerns in the Columbia River Basin, the chapter shows how data inputs could be transitioned from a relational to a graph database in an attempt to expand modes of digital analysis in environmental decision-making. Ultimately, insights derived from digital data are reflections of which inputs are encoded and how, as well as the extent to which researchers and databases can speak a common language to answer creative questions. Shifting data structure is one potential site for reconfiguring what we might ask and thereby come to understand about nature through digital data.

Data Structures

To systematically extract meaning from data, we—or the programming languages we have developed—rely on data that is consistently structured. Data structures define the types and properties of data that can be input, as well as the methods by which that data can later be analyzed, summarized, or extracted. Different data structures can change the underlying patterns and encoding of data in a database.[5] Perhaps most important, the structure of a database can exclude or simplify information that does not conform to its predefined standards. Examples include truncating long strings of text, the inability to store multimedia such as video, and limitations on the size and accuracy of numbers. As environmental decision-making continues to rely on digital tools and data stores, these structures will be a part of translating complex worlds into data. If we take the structure of databases for granted, we miss seeing the work that these frameworks do in shaping the information on which later analysis is contingent.

The case study presented in this chapter focuses on the process and outcomes of shifting between two different data structures: relational and graph. Data with spatial attributes typically conforms to the relational data structure necessary for use in most geographic information system (GIS) software. Increasingly, these underlying databases are being removed from users' view and inspection with server-side processing, but they remain in use. This chapter considers the shifts in measurement and value possible through the use of a graph data structure, as opposed to the relational model typically relied on to store attributes of place.

Early GIS developed alongside the emergence of the relational data structure in the late 1960s and early 1970s.[6] The novel development in the first GIS was the ability to store and query spatial information; this continues to define the difference between a GIS and other computational data environments. While a variety of spatial data models and structures are incorporated into GIS (including graphs), nonspatial attributes are exclusively stored using a relational data model that consists of data organized into tables. Each individual entry or item of data is stored as a row in the database and indexed with keys, or identifiers, to allow for joining multiple tables (table 3). A relational database structure has the benefit of being accessible to a wide range of analytical skill sets and software applications—for bookkeeping or tracking a phenomenon over time, for example. However, relational databases are also rigid in form and inefficient for large datasets compared with more recently developed alternatives, including graph databases.

TABLE 3. Relational database structure

	Identifier / ID	Attribute	Attribute
[Data Item 1] —>	Value	Value	Value
[Data Item 2] —>	Value	Value	Value

Source: Created by author.

In contrast to a relational data structure, a graph data structure is based on nodes and edges, where relationships between data items are stored directly for each entry, not in external schema (fig. 22). Graph databases allow for a wide variety of data types and massive amounts of data to be included without compromising query efficiency. Query languages designed for graph databases are both efficient and intuitive: by centering relationships, queries can follow lines of questioning without incorporating arbitrary keys. A graph data structure makes data types flexible and scalable, opening up opportunities to recode or uncode nature in our databases.

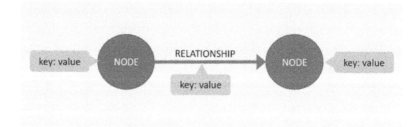

22. Directional graph data structure. Courtesy of author.

Coding Land and Water as Resources

Throughout the histories of colonialism, maps have played key roles in both consolidating state power and facilitating its continued expansion onto—and investment in—resource-rich lands.[7] Maps' simplifications of complex landscapes implicitly reproduce dominant structures of value and exclude alternative narratives by limiting what is made visible. Continuing this tradition, geospatial databases assembled for resource mapping remain constrained by the political-economic systems on which they are built. Whether abstracting agricultural yield, forest productivity, ecosystem service flows, or biodiversity "hot spots," datasets that aim to translate nature into investable assets focus on only a subset of values and obscure the actually existing actors, relations, and messiness of spaces.[8] As our increasing reliance on automated or technical production of knowledge coincides with climate crisis and growing resource scarcity, questioning who is valued and whose values matter in digital representations of complex natures is critical—otherwise inequitable trade-offs are all but certain.[9] Data-driven land use decisions, such as where to increase biodiversity protections or site land-intensive renewable energy projects, risk excluding the values of communities and ecosystems most affected by physical or legislative changes on the landscape. Some national-scale datasets, for instance, depict large swaths of land as empty and open to investment, diminishing local residents' land claims and setting the stage for continued land dispossession.[10]

Regardless of underlying "green" intentions, the expansion of renewable energy will necessitate land use trade-offs, resulting in uneven socie-

Armistead

tal burdens.[11] But who will be burdened most? What will this transition look like? Where will it take place? These questions are far from settled.[12] Databases and maps of renewable energy potential therefore are a timely site for contesting and reconfiguring the inscription of value in digital representations of nature. Like land, fresh water is shaped by entrenched power structures and often abstracted from its social context to appear available, for instance, for dam building to generate renewable hydropower energy.[13] Analysis of hydropower potential mapping brings together individual and community-level relationships with rivers with a broader framing of land and resource extraction.

The New Stream-Reach Development Resource Assessment

This case study focuses on the New Stream-Reach Development Resource Assessment (hereafter the NSD assessment), a project mapping the hydropower potential of currently undammed rivers and streams in the continental United States. A component of the National Hydropower Asset Assessment Program (NHAAP), it was completed in 2014 and funded by the U.S. Department of Energy.[14] The NSD assessment contains two broad categories of data: the inputs necessary to calculate hydropower potential per stream reach in megawatts and aggregated environmental factors to indicate risks involved with dam building.

First, I consider NSD assessment environmental attributes and their current depiction in a relational database. For the purposes of the NSD assessment, environmental attributes consist of ecological, socioeconomic, legal, or geopolitical concerns.[15] All data inputs are available for download from the project's web page, aggregated to watersheds at HUC08 scale.[16] This data dissemination aims to meet the project's goal "to produce datasets and tools that allow for multiple analyses to be conducted by different organizations and individuals using a wide variety of development scenarios and assumptions."[17] But what are some of these possible wide-ranging analyses? How might the database itself be constraining how information is placed into and later extracted from this aggregation of environmental attributes?

Based on the NSD assessment, 32 percent of U.S. hydropower potential—untapped energy of rivers—exists in the Pacific Northwest, made up of the Columbia River Basin and the smaller coastal

23. Map displayed in Kao et al., *New Stream-Reach Development.*

watersheds of western Washington and Oregon.[18] This is the largest energy potential of any hydrologic region in the continental United States (see fig. 23). At the same time, the NSD's aggregated environmental inputs show a high percentage of stream reaches in this region with ecological, social, and legal concerns. While dams and cheap power have shaped the recent growth of the Pacific Northwest, today the hurdles of gaining public support and government permits are extensive. These tensions between hydropower potential and concern, coupled with deep roots of settler colonialism and existing hydropower production, make the Columbia River Basin an engaging site for thinking through the ways data structure shapes the inputs and outputs of digital depictions of nature and society.

The Big River

Over the last two hundred years Northwest tribes and Canadian First Nations have witnessed erosion of livelihoods, disregard for shared values, and silencing of concerns, all in the name of progress, stability, and economic growth in the Columbia River Basin.[19] When Indigenous communities alone inhabited the area, arriving settlers saw the land

Armistead

as empty and its rivers open to opportunity. These colonizers quickly made the river work for them by claiming the powerful waters for transportation, hydropower, agriculture, recreation, and other economic opportunities.[20] Yet despite extensive changes in land and water use in the Northwest, Indigenous communities—entwined in the cycles of river systems from time immemorial—remain intimately involved with the region's rivers. The foundational relationships of Northwest tribes with Wimahl or Nich'i-Wàna (the Big River, also known as the Columbia River) must be recognized in resource decisions.[21]

In the Columbia River Basin many historic and ongoing debates over the river as a resource revolve around salmon—and salmon stories form the foundation of my exploration of graph data structures. As a "first food" and part of tribes' spiritual and cultural identity, salmon are integral to tribal members' sense of place, religious practices, sustenance, economic livelihood, and understanding of life and renewal.[22] Richard White describes the centrality of salmon—using the language of graph networks—by depicting pre–European conquest Indigenous peoples as "a society of dense networks of relations," in which "the salmon fisheries formed a basic node where the lines of human relationships intersected."[23] The Columbia River Inter-Tribal Fish Commission describes the importance of salmon to these peoples: "Salmon have shaped the culture of the newcomers to this region just as they shaped tribal cultures before them. Salmon are the icon of this place. They are valued as food, as a resource, and as a representation of the wildness and wilderness for which the Pacific Northwest is known. They shape our land use policies and power grid. Whether they realize it or not, every single person in the Northwest is a Wy-Kan-Ush-Pum. We are all Salmon People."[24]

The focus on salmon presented here barely brushes the surface and does not aim to capture all values, particularly the cultural significance of these species. In the subsequent discussion, I foreground a graph data structure as one opportunity to represent these shared relations and create space for incorporating broader constructs of value through new data inputs and database connections. The example that follows is one of many stories of the Columbia River Basin to be told through this data and graph data structure. Connections among river

Environmental Inputs	Data Source	Data Type
critical habitats (no. species)	NatureServe digital distribution maps of freshwater fishes of the United States	Num (Species Count)
Endangered Species Act (ESA) federally listed fish species (no. species)	U.S. FWS endangered species program	Num (Species Count)
International Union for the Conservation of Nature (IUCN) species of concern (no. species)	NatureServe explorer species data	Num (Species Count)
potadromous or anadromous fish (no. species)	NatureServe explorer species data	Num (Species Count)
protected land (presence/absence)	USGS GAP analysis – Protected area database of the United States	Binary (presense/absence)
land-ownership index (no. entities)	USGS GAP analysis – Protected area database of the United States	Binary (presense/absence)
land-designation index (no. designations)	USGS GAP analysis – Protected area database of the United States	Num Count
U.S. national park (presence/absence)	USGS GAP analysis – Protected area database of the United States	Binary (presense/absence)
Wild and Scenic River (presence/absence)	USGS GAP analysis – Protected area database of the United States	Binary (presense/absence)
303d listed waterbodies (no. waterbodies)	U.S. EPA impaired waters and total maximum daily load	Num (number of waterbodies)
American Whitewater boating runs (no. boating runs)	American Whitewater, National Whitewater Inventory (AW, 2012)	Number of boat runs
boat ramps (no. boat ramps)	DeLorme Publishing Company (2012)	Number of access locations
fishing access points (no. access locations)	DeLorme Publishing Company (2012)	Number of access locations
surface water use (l/day-1 • km-2)	USGS Water Use in the United States	Num Continuous
ground water use (l/day-1 • km-2)	USGS Water Use in the United States	Num Continuous
urban land cover (%)	Multi-Resolution Land Characteristics Consortium [MRCL] via National Fish Habitat Action Plan [NFHAP]	Num Continuous – percentage
population density (individuals/km-2)	US Census via National Fish Habitat Action Plan [NFHAP]	Num Count
dams in local watershed (no. dams)	National Fish Habitat Action Plan [NFHAP]	Num Count
total dams in entire upstream network (no. dams)	National Fish Habitat Action Plan [NFHAP]	Num Count
land disturbance index	National Fish Habitat Action Plan [NFHAP]	Num Index

24. NSD environmental data inputs. Courtesy of author.

flows and water use, current energy infrastructure, populations, recreation opportunities, and other industries are all additional worthwhile investigations.

Data Inputs and Current Structure of Hydropower Potential

To standardize and share environmental attributes for U.S. stream reaches, the assessment team at the NHAAP processed data from a number of public sources. Figure 24 depicts the environmental inputs, their sources, and data types included in the national hydropower potential database. Even given this relatively long list and the underlying labor to construct these datasets, aggregated data for the NSD assessment represents only glimpses of the ecological, socioeconomic, and geopolitical contexts in the Columbia River Basin.

The framework of relational databases, along with constraints imposed by popular spatial data types, limits the amount and richness of data within an aggregated database such as the NSD assessment of environmental attributes.[25] Each dataset listed in figure 24 contains details that were simplified for use in hydropower potential mapping by the NSD assessment team. To construct a manageable output, complex data inputs were reduced to limited, predefined data types—namely, summary counts and binary (presence/absence) variables. In this process, the assessment team focused on distilling data inputs for simplicity and efficiency rather than retaining richness in the original larger, though still reductive, datasets. Database structure thereby limited the information translated from one digital dataset to another, to some degree influenced by the decision to store information in a relational database with a key-based schema.

For example, while including a count of endangered species is useful, knowing instead which species, particularly of anadromous fish, and their ranges and populations would be more useful. Indicating the presence or absence of protected areas does establish a baseline but overlooks the many categories of land cataloged in the U.S. Geological Survey's Protected Areas Database (PAD) with significantly different uses, from military installations and national parks to tribal reservations and private conservation lands. Population counts offer a general indication of the potential impact of planned hydropower development on surrounding communities, but including data about political leanings, primary industries, or outdoor recreation preferences would provide a much more nuanced understanding. In attempting to standardize and create aggregate indices, much of the potential richness of environmental and social data inputs, which is captured in the original digital sources, is erased.

Besides the limitations described, there are also some environmental factors that are simply left out. While any mapping will remain necessarily partial, certain inputs could be considered and would expand the analysis. Though the focus of this case study is not on expanding datasets per se, one missing characteristic of the landscape could not be overlooked. In perpetuating colonial power structures, the current NSD assessment environment inputs do not include lands ceded by tribes to the federal government. Even though reservations and other tribal-owned lands are included in PAD, they are not identifiable once protected lands are reduced to a count variable. Data on ceded territories was the only additional input I made to the graph data structure described in the next section—though many other environmental and social inputs may also provide value, such as species interdependencies, other population demographics, zoning ordinances, road infrastructure, geotagged photos, or community-generated data. Incorporating additional data is one way to shift values encoded in hydropower potential mapping. Because not all data types can be integrated into geospatial relational databases, a graph data structure offers increased flexibility for alternative data types such as multimedia files or long text blocks. Including these inputs would provide further opportunity to expand whose values are represented in digital representations of nature.

Transitioning to a Graph Data Structure

To test the potential of varying data structure, I constructed a new graph database from the same inputs as the NSD assessment, with the single addition of ceded territories. Where feasible, I used data inputs directly from their original source to avoid the aggregations of environmental inputs present in NSD assessment hydropower potential attributes. This process involved subjective decisions in preprocessing, particularly in determining which inputs would become individual nodes, as opposed to properties of other nodes, and what would delimit my categorization of relationship types. Preparing and loading data into a graph database also required decisions on the types and properties of relationships between various data. For example, stream reaches were intuitively "upstream" or "downstream" from one another based on flow direction, but I categorized towns and cities as either "along" a river reach or "near" it with a relationship of distance for weighting.[26]

In addition to decisions on relationship type, limits were set on the geographic extent at which connections would be mapped. When processing protected areas, for example, the distance from each stream reach to every protected area in the United States could have been calculated and included in the database; this level of detail went beyond the needs of my inquiries but could benefit projects focusing on patterns of land ownership or management. I instead included all protected areas within twenty miles of proposed dam locations—an arbitrary cutoff but one retaining flexibility for analysis, since each protected area was added to the graph database as a node with its individual attributes (e.g., acreage, management agency) and was related to stream reaches by distance. The graph database was expanded to accommodate further data, particularly the relationships between stream reaches and ceded tribal lands. Indigenous land claims are a critical input to include if these mappings are to begin breaking down colonial representations of space and contest U.S. tribes' underrepresentation in resource use decisions.

Traversing the Graph with Salmon Stories

Once NSD assessment environmental inputs and Indigenous territories are stored in a graph database, what lines of questioning could test the opportunities of this alternate data structure? In the example that follows, I focus on hydropower development impacts on salmon and Indigenous communities, topics central to water resource debates in the Pacific Northwest. Many other interconnected questions and analyses could be explored through this database, including impacts to recreation, agriculture, endangered species, and the local community more broadly. The query examples presented in this section engage with one hypothetical question: Considering a new dam site and hydropower facility that is known to exacerbate declining upstream salmon runs, which U.S. tribes should be consulted regarding the loss of this culturally significant species? Reducing the complex landscapes of the Columbia River Basin to digital geospatial data limits what can be known through the data; defining a data structure places further constraints on future analysis. Even a graph database representing rivers, populations, infrastructures, and nonhuman species is inscribed with values and contingencies before results are ever drawn from it. Accepting these constraints, I first set out to determine salmon impact extent: which salmon populations will be affected and where within the river system these fish move during their life cycle.

Here is a code snippet of Cypher, a declarative query language for graph databases.[27] This nested query follows the foundational structure of Cypher, *()–[:consists of]–>()*, where a subset of nodes, or data items, are identified based on a particular relationship to another set of nodes.

```
MATCH c = (:Reach {HUC: "170602090802"})-[:UPSTREAM*]
->(n:Reach)

MATCH d = (n)-[h:HABITAT]-(:Species{name: 'Chinook
salmon'})

RETURN n
```

Starting with the watershed of the hypothetical hydropower dam site (HUC #170602090802), the first query line identifies all stream reaches upstream of the dam site. The second query line draws out only those upstream reaches with chinook salmon habitat present, as some upstream reaches may have natural or artificial barriers rendering them inaccessible to migrating salmon. With these two expressions I have identified an upstream impact extent based on the physical landscape and known life cycle of anadromous fish, as opposed to an arbitrarily defined impact distance.

Similarly, based on existing knowledge of a salmon's migration from spawning reaches to the ocean and back, downstream impact areas can be extracted from the graph based on its network structure. As is the case with all watersheds, tributaries eventually converge to a single river flowing downstream to the next watershed or to the ocean. Given the known exit point—from the Columbia River into the Pacific Ocean—I generated a direct path with the following query, returning only stream reaches directly between the proposed dam location and the Pacific Ocean, as opposed to pulling in all downstream anadromous stream reaches.

```
MATCH p = (:Reach {HUC: "170402190903"})-[:DOWNSTREAM*]
->(:Reach {HUC: "170800060500"})

RETURN p
```

So far in this analysis, the river reaches identified could be found using GIS software, with "routing" network analysis tools. Though not visible in the software's user interface, spatial features (in this case river segments) are stored as a graph—more specifically, a network capable of retaining relationship properties between nodes. However, in currently available GIS software, network tools run through separate windows and are one step removed from contextual attribute information that remains locked in a relational database schema. The following query depicts the intuitive flow of querying a graph across both spatial relationships and nonspatial attributes, identifying which Indigenous communities have ties to waters affected by proposed dam construction.

MATCH c = (:Reach {HUC: "170602090802"})-[:UPSTREAM*]->(n:Reach)

MATCH d = (n)-[h:HABITAT]-(:Species{name: 'Chinook salmon'}), (n)-[:WITHIN]-(y:TribeArea)

MATCH (y)-[:TRIBES]-(i:TribeName)

RETURN DISTINCT i.name

This Cypher query returns only the first occurrence of each tribe whose territory overlaps with affected stream reaches providing habitat for chinook salmon at any point during their life cycle. The different process of data extraction and the centrality of relationships in a graph data structure creates an alternative, flexible framework for asking questions about this ecological and social data. Additionally, more detailed and varied data can be added to the graph database without compromising usability for the end user.

Relying on known current tribal headquarters for Indigenous communities in the Pacific Northwest, I created a map representing the tribes that should be consulted if a new dam were proposed in western Idaho, based on queries of the graph database (fig. 25). As this visualization makes clear, a predefined impact area would fail to capture the relationships—between dam locations and salmon populations, and between salmon and tribes—as those affected are not distributed evenly across space. Relying on a graph database to begin to answer which Northwest tribes should be consulted when salmon are at risk of impact, the central focus becomes the relationships of flowing water, migrating species, and displaced peoples. A richer picture of the land-

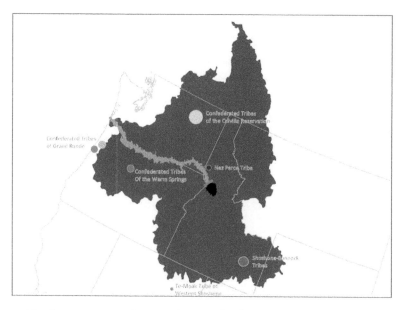

25. Map depicting extents of queries. Courtesy of author.

scape is created where different questions and more expressions of value may be possible, as opposed to counting attributes.

Conclusions

There are many continuing debates around flows of water, fish, food, power, and people in the Columbia River Basin that involve similar database tools to the N S D assessment to inform decision-making. In ongoing multinational renegotiation of the Columbia River Treaty, for example, tension exists between some tribes' desire for a treaty centered around ecosystem-based function and the U.S. federal government's interests in hydropower and flood protection.[28] Given the market-oriented modes of measurement inscribed in digital data, the case for hydropower and flood protection is straightforward, whereas quantifying the value of ecosystem-based function remains elusive, forcing the tribes to justify their approach within normative frameworks or risk appearing nonessential before treaty discussions even begin. On a stream-reach scale, funding for river restoration to improve salmon habitat or recreation opportunities is allocated throughout the basin every year based on cost-efficiency data and predicted environmental

Armistead

outcomes. What if our digital depictions of nature allowed for more flexibility in measurement and assessment? This is only one of many sites to contest resource use decisions in the Columbia River Basin, but critically thinking through choices of data structure is one step toward the larger project of broadening narratives of value surrounding the basin's ecosystems.

Digital tools, particularly those linked with geospatial datasets, that continue to advance inflexible assessments of nature bind us into ways of seeing constructed by settler colonial and free-market-centric value schemes. From influencing the data collection process to constraining the types of questions asked, the structure of databases incorporated into our tools, especially GIS, affects how and what knowledge is produced. Highlighting this work that data structures do opens up space for consideration of alternatives. The process of constructing and querying a graph database is one opportunity for data structure to broaden analyses of nature in hydropower potential mapping or other cases. Graphs place fewer constraints on data types, allowing for the incorporation of diverse inputs with different properties. Additionally, the ability of graph databases to efficiently sift through massive amounts of data eliminates the need to create index values or aggregates; the full richness of attributes can be included without compromising database usability. Finally, where relational databases assume the discreteness and uniqueness of data points (a stand-alone row for each entry), graph databases center around relationships—and this can change both the questions asked and the interactions with the data. By including alternative values in database structures for data-driven maps, opportunities expand for analyses and visualizations that take advantage of broader spaces for land and hydroscapes to become abstracted, measured, and compared.

Notes

1. Here I use *structure* to refer to the presence of any data model, not to call attention to the popular industry classification of relational databases as structured versus all others as unstructured.

2. Kitchin and Lauriault, "Digital Data and Data Infrastructures."

3. Robertson, "Nature That Capital Can See."

4. Burow, Brock, and Dove, "Unsettling the Land."

5. Peuquet, "Conceptual Framework and Comparison."

6. Goodchild, "Reimagining the History of GIS."

7. Pickles, *History of Spaces*.

8. For detailed characterizations, see Cotula, "International Political Economy"; Fairhead, Leach, and Scoones, "Green Grabbing"; Goldstein and Yates, "Introduction"; Li, "What Is Land?"

9. Norman, "Who's Counting?"

10. In "Visualizing New Political Ecologies," McCarthy and Thatcher demonstrate this reality in emerging datasets that facilitate large-scale mapping of renewable energy potential in Vietnam, specifically land-intensive wind and solar installations.

11. Huber, "Theorizing Energy Geographies."

12. Bridge et al., "Geographies of Energy Transition."

13. For details, refer to Mathur and Mulwafu, "Colonialism and Its Legacies"; Swyngedouw, "Political Economy"; Boelens et al., "Hydrosocial Territories"; Menga, "Hydropolis."

14. Kao et al., *New Stream-Reach Development*.

15. Kao et al., *New Stream-Reach Development*.

16. Oak Ridge National Laboratory, https://hydrosource.ornl.gov/dataset/hydropower-potential-new-stream-reach-development-conterminous-united-states.

17. Hadjerioua et al., *Assessment of Energy Potential*, xi.

18. Kao et al., *New Stream-Reach Development*.

19. White, *Organic Machine*.

20. White, *Organic Machine*.

21. Cohen and Norman, "Renegotiating the Columbia River Treaty."

22. CRITFC, "Tribal Salmon Culture." First foods are seasonal, wild edible species, such as salmon, game, and berries, that have long provided sustenance for Indigenous groups.

23. White, *Organic Machine*, 21.

24. CRITFC, "We Are All Salmon People." The Columbia River Inter-Tribal Fish Commission represents the Yakama, Warm Springs, Umatilla, and Nez Perce tribes and focuses on management policy and fisheries technical services that support the commission's mission "to ensure a unified voice in the overall management of the fishery resources, and as managers, to protect reserved treaty rights through the exercise of the inherent sovereign powers of the tribes."

25. For example, text fields have limited length when associated with shapefiles, a common spatial data type.

26. Stream reaches also contained relationships to properties such as potential hydropower generation, species presence, and water use. These items were modeled as nodes to facilitate cross comparisons or connections between variables. After establishing these relationships, cleaning, and preprocessing, the data was loaded into Neo4j in CSV format using Cypher queries (Neo4j's graph query language).

27. Cypher is an open-source declarative query language developed for Neo4j, which as of 2022 consists of both open-source and proprietary versions. See https://neo4j.com/open-source-project/.

28. See the Upper Columbia United Tribes (UCUT) references on foregrounding

ecosystem-based function in the Columbia River Treaty at https://ucut.org/water/columbia
-river-treaty/.

Bibliography

Boelens, Rutgerd, Jaime Hoogesteger, Erik Swyngedouw, Jeroen Vos, and Philippus Wester. "Hydrosocial Territories: A Political Ecology Perspective." *Water International* 41, no. 1 (2016):1–14.

Bridge, Gavin, Stefan Bouzarovski, Michael Bradshaw, and Nick Eyre. "Geographies of Energy Transition: Space, Place and the Low-Carbon Economy." *Energy Policy* 53 (2013): 331–40.

Burow, Paul B., Samara Brock, and Michael R. Dove. "Unsettling the Land: Indigeneity, Ontology, and Hybridity in Settler Colonialism." *Environment and Society* 9, no. 1 (2018): 57–74.

Cohen, Alice, and Emma S. Norman. "Renegotiating the Columbia River Treaty: Transboundary Governance and Indigenous Rights." *Global Environmental Politics* 18, no. 4 (2018): 4–24.

Cotula, Lorenzo. "The International Political Economy of the Global Land Rush: A Critical Appraisal of Trends, Scale, Geography and Drivers." *Journal of Peasant Studies* 39, no. 3–4 (2012): 649–80.

CRITFC (Columbia River Inter-Tribal Fish Commission). "Tribal Salmon Culture." Accessed May 5, 2019. https://www.critfc.org/salmon-culture/tribal-salmon-culture/.

———. "We Are All Salmon People." Accessed May 5, 2019. https://www.critfc.org/salmon-culture/we-are-all-salmon-people/.

Fairhead, James, Melissa Leach, and Ian Scoones. "Green Grabbing: A New Appropriation of Nature?" *Journal of Peasant Studies* 39, no. 2 (2012): 237–61.

Goldstein, Jenny, and Julian Yates. "Introduction: Rendering Land Investable." *Geoforum* 82 (2017): 209–11.

Goodchild, Michael F. "Reimagining the History of GIS." *Annals of GIS* 24, no. 1 (2018): 1–8.

Hadjerioua, Boualem, Shih-Chieh Kao, Ryan A. McManamay, M. Fayzul K. Pasha, Dilruba Yeasmin, Abdoul A. Oubeidillah, Nicole M. Samu et al. *An Assessment of Energy Potential from New Stream-Reach Development in the United States: Initial Report on Methodology, Technical Manual 2012/298*. Oak Ridge TN: Oak Ridge National Laboratory, 2013.

Huber, Matt. "Theorizing Energy Geographies." *Geography Compass* 9, no. 6 (2015): 327–38.

Kao, Shih-Chieh, Ryan A. McManamay, Kevin M. Stewart, Nicole M. Samu, Boualem Hadjerioua, Scott T. DeNeale, Dilruba Yeasmin, M. Fayzul K. Pasha, Abdoul A. Oubeidillah, and Brennan T. Smith. *New Stream-Reach Development: A Comprehensive Assessment of Hydropower Energy Potential in the United States*. GPO DOE/EE-1063. Wind and Water Power Program. Washington DC: Department of Energy, 2014.

Kitchin, Rob, and Tracey Lauriault. "Digital Data and Data Infrastructures." In *Digital Geographies*, edited by James Ash, Rob Kitchin, and Agnieszka Leszczynski, 83–94. London: sage, 2018.

Li, Tania M. "What Is Land? Assembling a Resource for Global Investment." *Transactions of the Institute of British Geographers* 39, no. 4 (2014): 589–602.

Mathur, Chandana, and Wapulumuka Mulwafu. "Colonialism and Its Legacies, as Reflected in Water, Incorporating a View from Malawi." *Wiley Interdisciplinary Reviews: Water* 5, no. 4 (2018): e1287.

McCarthy, James, and Jim Thatcher. "Visualizing New Political Ecologies: A Critical Data Studies Analysis of the World Bank's Renewable Energy Resource Mapping Initiative." *Geoforum* 102 (2019): 242–54.

Menga, Filippo. "Hydropolis: Reinterpreting the Polis in Water Politics." *Political Geography* 60 (2017): 100–109.

Norman, Emma S. "Who's Counting? Spatial Politics, Ecocolonisation and the Politics of Calculation in Boundary Bay." *Area* 45, no. 2 (2013): 179–87.

Peuquet, Donna J. "A Conceptual Framework and Comparison of Spatial Data Models." *Cartographica* 21, no. 4 (1984): 66–113.

Pickles, John. *A History of Spaces: Cartographic Reason, Mapping, and the Geo-coded World.* Hove, England: Psychology Press, 2004.

Robertson, Morgan M. "The Nature That Capital Can See: Science, State, and Market in the Commodification of Ecosystem Services." *Environment and Planning D: Society and Space* 24, no. 3 (2006): 367–87.

Swyngedouw, Erik. "The Political Economy and Political Ecology of the Hydro-social Cycle." *Journal of Contemporary Water Research & Education* 142, no. 1 (2009): 56–60.

White, Richard. *The Organic Machine: The Remaking of the Columbia River.* New York: Hill and Wang, 1995.

··

FIFTEEN

How Forest Became Data

THE REMAKING OF GROUND-TRUTH IN INDONESIA

Cindy Lin

O n April 15, 2020, Microsoft president Brad Smith launched the Planetary Computer, a computing infrastructure that aims to aggregate data on biodiversity, water tables, forestry data, carbon, and waste across the world to provide a searchable dataset. Smith elaborated that users could now exploit machine-learning tools to conduct environmental assessments that are "faster, cheaper, and—for the first time—operate at a truly global scale." He continued, "It should be as easy for anyone in the world to search the state of the planet as it is to search the internet for driving directions or dining options."[1] The analogy between searching the planet and searching for restaurants suggests that nature, once transformed into digital data and processed for real-time distribution, becomes a highly predictable system—otherwise already captured in the term "ecosystem," often described as "components operating in chains of cause and effect."[2] To translate the planet into data implies that humans can better manage and control the world, if only it were rendered computable.

As Big Tech promotes machine learning and aggregated datasets as a solution for our climate crisis, Indonesia's president Joko Widodo issued Presidential Regulation No. 39 on One Data (Satu Data in the Indonesian language) in 2019, an initiative to build a similar data infrastructure that monitors and analyzes environmental changes. One Data aims to improve the interoperability and standardization of government data—from agriculture and mining data to demographics and financial debt—for policy-making and economic planning in Indonesia.[3] Satu Data also encompasses the One Map Policy, a national project to remap 18,309 islands in an attempt to further infrastructure and resource extraction projects in rural Indonesia. Implemented by

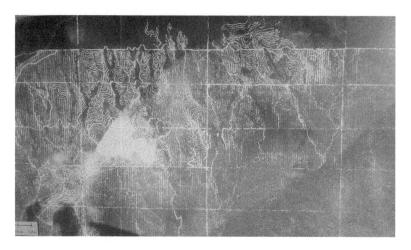

26. A satellite image used by data technicians to classify palm oil plantations in Kalimantan. Courtesy of author.

private-sector geospatial companies, One Data integrates geospatial data from One Map with other key statistics to make complex social and environmental problems transparent and accessible through trends, patterns, and correlations. What does it mean when tech firms and corporations are as involved in state planning and decision-making as bureaucrats are?

At the heart of the Planetary Computer and the One Data project are visions not only to map the whole planet but also to gain insights from the volume of data that burden bureaucratic decision-making. In bureaucracy data-driven technologies promise to assist administrators in identifying and preempting high-risk events such as fire, crime, economic trends, and water and land disputes. Accordingly, a political ecology of data enables one to question exactly how and whose truths are made possible through data infrastructures like One Data and One Map. Maps, after all, detail more than the biophysical constitution of territory; they reveal the compulsions, practices, and desires for accuracy and measurement. To make a planet computable not only gives one the power to see, collect, or survey a vast landscape but also distills something down for action. A computable planet becomes an object for corporations or government to act on, or as Jennifer Gabrys puts it, "a medium of stabilized operation and control."[4]

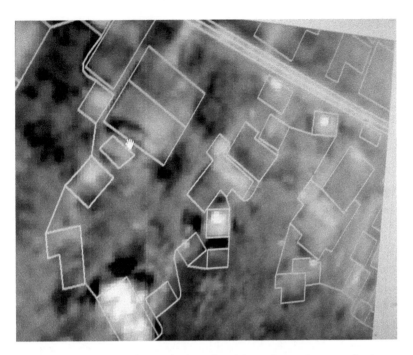

27. Munir attempting to redraw multiple roofs with limited time. Courtesy of author.

For a developmentalist postcolonial state like Indonesia, state leaders have similarly placed special emphasis on how technologies can construct a stabilized national identity. As historian of technology Suzanne Moon argues, in postcolonial Indonesia, technology and national identity mutually reinforce one another. Technologies produce "materially compelling interpretations of the character of Indonesia expressed in pursuits of modernity, even if these technologies can be impractical and out of reach."[5] Less remarked on, however, are the practices and aspirations of those who make the very data technologies used to stabilize and manage the planet from Indonesia. I argue that Indonesia's aspiration to remap its islands segregates and divides who gets to *see* and who gets to *know* nature along class and racial lines.[6] More important, the remapping effort positions new mapping technologies and techniques as proxies of transparency, ensuring unmediated access to events happening in Indonesia's forest, out there—and out of sight (fig. 26).

In Pursuit of a Single Map

Munir is a high school–educated data technician hired by a private geospatial mapping company in Bandung, an urban provincial capital on the island of Java, to produce topographic maps. He is one of a team of technicians working under contracts from the National Mapping Agency to draw squares and rectangles around vast swaths of buildings and property across the archipelago. Sitting with Munir, I noticed him staring at pixels instead of satellite images. Working at a scale of 1:25,000, Munir used his mouse to quickly click and draw a perimeter around what I learned were roofs (fig. 27). The roofs are in Temajuk, a coastal city on the island of Kalimantan, a region that has witnessed the world's fastest forest clearing rates since 2012 as the oil palm sector has expanded across rural Indonesia.

I asked Munir, "Why not zoom in?" thinking this would ensure his clicks were made with more precision so that nothing was missed. He told me, "My boss wants us to work fast." He then explained that zooming in would give him and his colleagues too much content to work with. Even if state agency scientists demand that every building be drawn, his supervisors were afraid that the technicians would interpret too much. Every minute spent in their office facilities, on their computers, and using software counts for this pool of technicians, paid far too little to meet the agency's ambition for an archipelago to be remapped.

During regular three-month-long field surveys to plantation frontiers like Temajuk, Munir and his colleagues cross-reference the maps they plotted in the office. The exploited terrain poses a danger to the visiting technicians as they tread on eroded and restricted hills, succumb to dislocated knees, and negotiate threats of death and injury. Field surveyors, pressed to verify a target number of maps within a limited period of time, often must camp for several nights in remote forested areas, without food, water, or fuel or signal to return to their base camps. This work is not easy, and most are paid only 3 million rupiah (about US$200) monthly, a quarter of the average income in Bandung. Munir shrugs when I ask about his wages, reasoning that maps are crucial for Indonesia's *pembangunan* (development), an ever-shifting ideology that has haunted the nation since its plans for rice self-sufficiency began in the 1950s.[7] Yet Munir and his fellow technicians insist that their deskwork and field surveys as direct participants in state forestry protection provide them, migrant lower-class men from rural Indonesia, with opportunity—a social mobility otherwise out of reach.

The contract employing the technicians stemmed from an incident in December 2010, when former Indonesian president Susilo Bambang Yudhoyono compared two conflicting maps of Papua Island. The two forest maps, one published by the Ministry of Forestry and the other by the Ministry of Environment, were presented in a cabinet meeting regarding a nationwide ban on plantation permits in primary forests. Each map showed primary forests of different sizes and with different boundaries. Researchers attributed this information asymmetry to different forest classifications and mapping methodologies practiced by the two ministries.[8] A scandal broke out, with environmentalists claiming that these discrepancies were an example of how corrupt officials manipulate maps and issue plantation permits on protected forestland.

Beginning in 1998 Indonesia's timber and oil palm industry became increasingly decentralized and privatized.[9] Over time resource extraction became difficult to monitor, with central and provincial offices producing multiple contradictory maps.[10] Subsequent allegations of inaccurate maps and weak law enforcement by international aid donors and foreign governments led to broader uncertainties over forest stocks. This deterred foreign investment in Indonesia's emerging carbon reduction schemes—schemes that required new standards of transparency when estimating forest loss. One such scheme is the Reducing Emissions from Deforestation and Forest Degradation (REDD+) program, a United Nations–led global climate initiative that treats forest carbon as a metric for financing a country's forest conservation efforts.[11] The promise that Indonesia's export trade could be bolstered by new forms of climate finance renewed focus on transparency in mapping forest loss.

Under a 2016 presidential decree, Indonesia's National Mapping Agency employed teams of Earth scientists and geospatial data technicians to accelerate the remapping of Indonesia under the One Map Policy so that a single base map would standardize how different ministries classified forest cover.[12] The base map, conventionally understood as a ground-truth to physical reality, was proposed as the solution to know where exactly forests have been cleared. Since 2016 state scientists have implemented a suite of artificial intelligence and remote sensors to expedite the remapping process. These new mapping technologies are aimed at removing human interpretation, with state scientists viewing the existing industry of geospatial data technicians as costly, inefficient, and vulnerable to bribes. At the core of this remapping effort is the administration's belief that maps should resemble the terrain they seek to represent.

Yet the National Mapping Agency's senior bureaucrats have always been aware that maps could never truly adhere to the terrain. For instance, the agency's deputy director has stated to the Indonesian press that data on palm oil plantation ownership should be kept confidential.[13] Herein lies a tension between bureaucratic desires to conceal deforestation and the official appearance of attempts to reveal what is happening on the ground. The accuracy of forest maps is not only due to technical expertise or the practice of transparency. A map's legibility can also be understood as part of a measured and managed public revelation—and concurrent concealment—of information, narratives, and images of trees that still stand.

Michael Taussig, developing Elias Canetti's dictum that secrecy is at the heart of power, argues that social orders are based on "public secrets": forms of knowledge that are generally known but must not be overtly acknowledged.[14] Simply put, one has to know what not to know. Public life, its discourse, and practice, then, depend on the management of transgression.[15] In Indonesia it was understood that existing forest maps contained inaccuracies. From the fixed borders that enabled industrial plantation expansion to inconsistent mapping standards, state maps were dissimilar all the way down. The recent remapping effort is an attempt at undoing such legacies, even when Indonesia's state legitimacy in forest management relies on the maintenance of what is already known but cannot be said: no map can truly capture all existing forestland. As I see it, state officials envision One Map as a political means not only to stabilize national identity but also, as I show, to experiment with new ways to reveal and make transparent the country's forest resources.

The pursuit of different versions of correspondence between terrain and map segregates who gets to see and who gets to know what makes a forest. The labor of geospatial data technicians who are central to this operation to *see*—by constructing and calibrating base maps from satellite imagery, software, and in-person, in-the-field surveys—is viewed as an obstacle to *knowing* a truly accurate map. Senior bureaucrats characterize data technicians' labor as necessary but inefficient—a mere act of data collection rather than trained expertise. Bureaucrats have advanced data science projects to automate the delineation of forest borders—to know the forest in precise ways that crowd out what, for others, is actually there.

Locating Boundaries in the Office and Field

The operation of correspondence practiced by data technicians was made clear to me when I was taught "how to see" at the technical office in Bandung. Working through a montage of green and brown, Munir commented, "When the brown pixel changes into a green pixel, what you see is the edge of a forest and the beginnings of a house. You try your best to draw a square, but always a little forest or house will escape." His earlier trace was returned in red, marked as incorrect by the National Mapping Agency. "They want us to draw every house, but I can hardly see them without zooming in. And if I zoom in, I will never finish this map in time. I will never be paid."

Munir agonized over the deadlines set by the National Mapping Agency. He had spent hours determining what pixels made the cut. Perhaps drawing in this green pixel matches the vegetation nearby and adds to Indonesia's total forest cover count; perhaps adding a brown pixel grants more property for the house owner. The blur in the images makes for a deliberation that also causes his wrists and fingers to grow sore while deciding what should be included and excluded. At one point I asked about the interstice where a green pixel had ended. Unsure, he hovered his mouse back and forth over the shades of green that rendered into a murky brown. Indeed, "there are no straight lines" on the edges of Kalimantan's forest; technicians "feel" these borders in creating mapped representations.[16] All the while the technicians are reminded

28. Data technicians on a German logging concession in Kalimantan.
Courtesy of author.

that an extra pixel adds at least five meters to the total calculated area, posing a serious challenge to Indonesia's ministries in their bid not to misrepresent forest loss.

In 2017 I followed Munir on a field survey to search for a Global Positioning System (GPS) point plotted on a satellite image of West Kalimantan that had been taken earlier in 2015 (fig. 28). Halfway on our journey into the thick of forest dew and sludge, Munir's navigator stopped working. Local residents had earlier warned us that we were already trespassing into the concession areas of a German timber-logging company and would be in grave danger if spotted. Munir refused to halt his search and return to camp. "I have to find this point," he said. "The Mapping Agency wants to know if this is a forest or a field. I don't want to delay our team."

To speed up these remapping activities and avoid the contentious and costly field surveys, I later learned that Munir's colleagues were instructed to classify forests using light detection and ranging (lidar) data from the National Mapping Agency. Lidar has three features: intensity values (the amount of light energy recorded), elevation values, and a return number (the total number of returns for a given laser pulse) recorded from a remote-sensing device fitted to an airplane or satellite. These values are compiled into a 3D data "point cloud" that

29. Lidar point cloud of river and tropical forest in Kalimantan. The different colors represent the range of elevation values. Courtesy of author.

is unlike a flat pixelated satellite image, emphasizing features that cannot be observed by plain sight (fig. 29).

Further contrasting a point cloud with a pixel, an interlocutor at the National Mapping Agency told me that lidar can help technicians think less and map faster: "Unlike a point cloud, which intuition will tell you to join together based on the proximity of dots, a single pixel gives *too much* information. Pixels make an operator imagine too much." Munir's supervisor confirmed this, commenting that technicians' "hands and eyes are meant to work fast without thought." Instead of drawing lines around pixels to classify building and property reserve, technicians now join proximate dots. A reprogramming of efficient minds is set in motion by points.

Deep Learning Ground-Truth

Incremental efforts to limit technicians' expertise extend to the application of deep-learning techniques in state mapmaking contracts. Bayu, a university-educated Earth scientist from the National Mapping Agency in Jakarta, has developed a method to transform a 3D lidar point cloud into 2D pixel-based images. These images are what Bayu then uses to train a fully convolutional network (FCN) to classify point clouds into different features (e.g., trees and buildings). Ironically, while technicians allegedly see too much in pixels, Bayu's FCN model returns to pixels to categorize points. Extracted from lidar, these pixel images also act as the ground-truth for Bayu to calibrate future outputs from his model.

Bayu described the steps he took to train his FCN model and develop a ground-truth image:

1. Superimpose a one-by-one-meter pixel grid onto a lidar point cloud. The grid is meant to represent a pixel-based image. There will likely be more than one lidar point within a pixel.

2. In each pixel of the grid, choose the lidar point with the lowest value. Each lidar point contains a value calculated from (1) elevation, (2) intensity, (3) return number, and (4) height difference (i.e., the difference in elevation between a lidar point and neighboring points). It is assumed that lidar points with higher values represent nonground features such as vegetation and buildings.

3. Label the pixels containing low values as ground. Points that exceed the threshold of fifteen centimeters are labeled as nonground points. The final tagged image is called the ground-truth image.

4. Create the training dataset using an unlabeled image extracted from the lidar point cloud (from steps 1 and 2).

5. Train an FCN on the dataset. The output of the FCN is a predicted label for each pixel.

6. Compare the output to the ground-truth image. If the predicted labels ("ground" and "nonground") are different from the ground-truth image, change the weights and parameters in the FCN model until the output's labels correspond to the ground-truth image.

7. Transfer the predicted label of each pixel back to the original lidar point cloud.[17]

Earth scientists like Bayu do not conduct field surveys to verify their maps. Instead, verification involves comparing the model's output with the ground-truth image, which itself is a derivative from the same output data, producing what geographer Louise Amoore views as a model of what has become "normal" in the data.[18]

Bayu's model attempts to mimic how data technicians draw boundaries. Distinguishing ground from nonground at the level of every pixel, his model is considered by his superiors as more precise than how technicians classify pixels. Absent from Bayu's work, however, was a concern for the implications of correspondence. Unlike the data technicians, who deliberate on what pixels to include and exclude and venture into the field to calibrate their judgment, Bayu felt alienated from his model's decisions. "It's just a black box," he told me.

Models that seek out patterns like Bayu's deep-learning technique resemble what computing scholar Dan McQuillan calls "machinic Neoplatonism." Neoplatonic approaches to science aim to reveal the "hidden mathematical order in the world" and go against experience.[19] Bayu models what is out there with traces of experience—the pixels— that can be quantified and computed in scales that exceed human apprehension. As critical data studies scholar Os Keyes claims, what differentiates traditional science and these data science examples is how the latter does not generate theory.[20] While the former attempts to explain truth by assuming a distanced position, the latter arrives at truth without explanation. In Bayu's case a model is designed to be the optimal technician, yet it challenges what he could see and know. He can only verify the model's output value against the ground-truth image—nothing more, nothing less.

During one workshop Bayu realized that his FCN network had failed to classify features unique to Indonesia, in part because the model's training data was generated from urban northern Europe. He attempted to add more object classes to the ground-truth image. These included "Uncleared Land" and "Waste," categories informed by his living in Jakarta, a metropolis well known to be heavily polluted. I understood these revisions as akin to a field survey, in their own way. Like Munir, stuck on the German-owned logging concession and committed to finding his GPS point, Bayu was locating his own set of object classes to calibrate ground-truth. Bayu's computational practices, while different from a data technician's tactful vision, are in every way iterative and experimental.

To date the National Mapping Agency has resisted scaling up Bayu's model to remap Indonesia. Senior bureaucrats claim that they do not know how it works. They compare the accuracy of Bayu's model to the work of data technicians, arguing that his computational techniques cannot be accurate unless their outputs are verifiable in field surveys. Although often viewed as an exemplar to efficiently estimate forest loss, these algorithmic classifications were deemed inscrutable.

Conclusions

In the Indonesian state's pursuit of a single map of the archipelago, the labor of data technicians is both displaced and venerated, albeit indirectly, in the recurring recognition of the need for field surveys to verify ground-truth. When seeing the forest is translated into wage labor, state agencies regard the technicians' work as too inefficient because, like Munir, they see too much. Data science initiatives, on the other hand, draw borders without explanation to remove this interpretive labor. In opposition to the less-educated data technicians, this work resembles a "tiered" seeing—to borrow from feminist technoscience scholar Mitali Thakor—where "certain perceptions are afforded higher authority."[21] Trained to recognize points instead of pixels, Bayu's model privileges a specific mode of knowing that is preemptive, even if experiential by design. Preemption here operates by forestalling and foreclosing potential futures in favor of a computationally calculated output. Points are converted into labeled pixels and fed into models that in turn label points anew: points feed points. Unlike the field survey, with data science ground-truth is found within the image itself, not in the forest.

Indonesia's forests are seen and known by the use of different techniques. These techniques either institute or automate proxies of human interpretation in the pursuit of ground-truth and aim to resolve a crisis of legitimacy in forest management. To position the Ministry of Environment and Forestry as a transparent and uncorrupt authority, officials implemented mapping techniques that were based on a hierarchy of expertise and appealed to notions of data objectivity. Yet in keeping with the secrecy of how forests are seen and governed by the state, each of these techniques also cultivates a *willful not knowing*. Against the backdrop of the decades-long expropriation of Indigenous land by patronage networks of timber and palm oil firms and central and district officials, broader shifts to institute efficiency and automation

Lin

in mapping enable bureaucrats to relinquish their knowledge of such inconvenient truths.[22] These technical initiatives recast the mapping of Indonesia as a preemptive activity that deters other ways of seeing and knowing the forest, rather than a product of trained expertise. The task of *knowing what not to know* emerges not only from the discursive theater of public human affairs but also out of the contest between the design and deployment of various mapping systems, field surveys, and data infrastructures.

For these techniques to cultivate a willful not knowing, technology needs to be positioned as ensuring transparency in Indonesia's political affairs. In 1998, when Indonesia experienced two major events—massive forest fires and the Asian financial crisis—the economic gains under President Suharto's New Order Regime were quickly undone. Across the nation, the magnitude of the Asian financial crisis affected elite support for Suharto's military dictatorship, as students led demonstrations that led to his resignation in May 1998.[23] Student protestors argued that transparency would bring to light the political and economic corruption of the New Order Regime, including Suharto's concessions to state-owned enterprises and close business affiliates to the country's largest credit sources.[24]

When Suharto resigned, the increasingly deregulated media outlets exposed a wide variety of scandals through photographic images, ranging from acts of corruption committed by ruling elites to sex scandals between politicians and famous celebrities. Anthropologist Karen Strassler argues that these acts of public exposure were met with an earnest sense of political possibility that a new era of transparency and accountability was possible for the Indonesian public. This was articulated through what Strassler calls the "evidentiary reading of images," a way of "revealing political realities through an indexical reference to a prephotographic real."[25] Evidentiary reading of images positions technology as an agent of transparency, a system for tracking the origins of an image, ensuring unmediated access to otherwise undisclosed events and complete transmission of truths. In this way the National Mapping Agency's control over how contracted data technicians see, how they map borders, and what tools they use to reveal deforestation exemplified post–New Order desires to increase transparency in the forest sector—which led to uncovering numerous political scandals and controversies.[26]

Contemporary data infrastructures that enable data science practices challenge the place of evidentiary reading and its allegiance to transparency. Unlike field surveys conducted to authenticate state maps, the implementation of data science in mapping practices reinvents what counts as the field or the terrain by extension. Earth scientist Bayu uses data science to verify his modeled outputs by comparing these predictions with a ground-truth dataset instead of surveying the forest itself.[27] The ground-truth in turn is derived from a benchmark training dataset of northern European landscapes and Bayu's parameters for ground and nonground data points. Indonesia's forests, on the other hand, are what technicians encounter to verify their maps. Accordingly, data science relies on modes of verification that are not circumscribed to expose what is otherwise hidden or out of sight—such as the search of a GPS point in a forest. In this way Earth scientists like Bayu who practice data science do not find one "true" map but instead produce an output that statistically approximates a ground-truth dataset, rejecting any single moment of revelation crucial for fulfilling ideals of postreform transparency: "openness, visibility, and accessibility."[28] Therefore, I view senior officials' rejection of Bayu's FCN model as rooted in a concern for transparency in contemporary environmental governance. The operations of data science displace Indonesia's growing obsession with the open accounting and exposure of secrets in the immediate post-Suharto period. Transparency, an ideal that the National Mapping Agency aimed to achieve by instituting and standardizing how data technicians see and map the forest, is now hindered by what constitutes ground-truth in data science.

I see data science, then, as not simply a new form of environmental governance or a technique allowing officials to claim neutrality and objectivity. Rather, data science is a conduit for questioning the political norm of transparency, once primary for revealing corruption in postreform Indonesia. In a world of big data, truth and insights emerge when data is given an adequate visual and cartographic form. Data science practices that Bayu engages in are infrastructure born of a larger government rationality that must learn as much as possible about forests not only through older administrative technologies like censuses and cadastral surveys but also by capturing statistical correlations and digital patterns. The political ecology of data, then, pushes one to reexamine how representing nature generates new hierarchies of expertise, often at the expense of exacerbating class and racial inequalities. Moreover, using data as infrastructure for decision-making shapes political norms of transparency and the legitimacy of social institutions. In Indonesia, contemporary data infrastructures change how forest can be revealed in ways that challenge the legitimacy of central forestry leadership, especially when data science remakes what is considered ground-truth, point by point, line by line.

Acknowledgments

This essay documents field research conducted between 2016 and 2018 in Indonesia. I thank editors Jenny Goldstein and Eric Nost for their invitation and meticulous reading of this book chapter, as well as Lisa Nakamura, Juno Salazar Parreñas, Sareeta Amrute, Silvia Lindtner, and Andrew Moon for their generous support and wit in the original e-flux article. Ultimately, this essay has been made possible by the data technicians and scientists who kindly shared with me their drawings, surveys, and musings.

Notes

1. Smith, "Healthy Society."
2. Brain, "Environment," 153.
3. "Jokowi Signs Regulation."
4. Gabrys, "Becoming Planetary."
5. Moon, "Justice, Geography, and Steel," 254.
6. In "Techniques of Use," Silvia Lindtner and I discuss further how a seemingly "positive" ideal to make technology "useful"—that is, to build systems and devices that advance social and technological progress—masks various forms of violence and injustice such as colonial othering, racist exclusions, and exploitation.
7. Setiadarma, "Cultivating Pembangunan."
8. Since 2015 both ministries have merged to form the Ministry of Environment and Forestry. Astuti and McGregor, "Responding"; Shahab, *Indonesia*.
9. It has been argued that this restructuring enabled district mayors and village leaders, backed by political elites in the Ministry of Forestry, to finance their budgets by faking or fixing permits for resource companies. See Tsing, *Friction*; Li, *Will to Improve*; Peluso, "Whose Woods Are These?"

10. Astuti, "REDD+ Governmentality."

11. Astuti, "REDD+ Governmentality."

12. A base map is a layer with geographic information that serves as a background setting for other maps. It establishes the geometrical and orientation reference for the viewer of a thematic map.

13. Jong, "Indonesia Calls on Oil Palm Industry."

14. Taussig, *Defacement*, 7.

15. Mazzarella and Kaur, "Between Sedition and Seduction," 14.

16. Chao, "There Are No Straight Lines," 21; Ballestero, "Touching with Light," 14–17.

17. Repeat steps 2 and 3 to create an image of nonground points derived from selecting the highest values. Separate nonground points into finer classes: vegetation and building. Building and vegetation labels are created based on three assumptions: (1) a building's roof surface is composed of lidar points with similar elevation values; (2) a building's edges do not have neighboring points on all sides; and (3) the intensity value of vegetation is lower than that of buildings. Building and vegetation labels are transferred to nonground points from the result of the ground classification derived in step 3.

18. Amoore, "Doubt and the Algorithm."

19. McQuillan, "Data Science," 259, 261.

20. Keyes, "Gardener's Vision of Data."

21. Thakor, "Digital Apprehensions," 8.

22. Varkkey, "Patronage Politics."

23. Lee, "Styling the Revolution," 935.

24. Ibrahim, *Improvisational Islam*, 82–89.

25. Strassler, *Demanding Images*, 72–80.

26. Goldstein, "Knowing the Subterranean," 7–12.

27. Benchmark training datasets for machine learning are usually based on landscapes from the Global North. Open data initiatives such as Radiant Earth Foundation have attempted to better expand such benchmark training datasets by manually annotating and classifying high-resolution images of Global South landscapes, as documented by Radiant Earth, https://www.radiant.earth/strategic-plan/.

28. To give a crude analogy, verification in traditional mapping resembles opening a box to find something hidden, while automated mapping resembles circuitry with electricity that never leaks or leaves but continues to feed and generate a stronger circuit.

Bibliography

Amoore, Louise. "Doubt and the Algorithm: On the Partial Accounts of Machine Learning." *Theory, Culture & Society* 36, no. 6 (2019): 147–69.

Astuti, Rini, and Andrew McGregor. "Responding to the Green Economy: How REDD+ and the One Map Initiative Are Transforming Forest Governance in Indonesia." *Third World Quarterly* 36, no. 12 (2015): 2273–93.

Astuti, Rini Yuni. "REDD+ Governmentality: Governing Forest, Land, and Forest Peoples in Indonesia." PhD diss., Victoria University of Wellington, 2016.

Ballestero, Andrea. "Touching with Light, or, How Texture Recasts the Sensing of Underground Water." *Science, Technology, & Human Values* 44, no. 5 (2019): 1–24.

Brain, Tega. "The Environment Is Not a System." *A Peer-Reviewed Journal About* 7, no. 1 (2018): 152–65.

Chao, Sophie. "'There Are No Straight Lines in Nature': Making Living Maps in West Papua." *Anthropology Now* 9, no. 1 (2017): 16–33.

Gabrys, Jennifer. "Becoming Planetary." *Accumulation* (October 2018). e-flux Architecture. https://www.e-flux.com/architecture/accumulation/217051/becoming-planetary/.

Goldstein, Jenny E. "Knowing the Subterranean: Land Grabbing, Oil Palm, and Divergent Expertise in Indonesia's Peat Soil." *Environment and Planning A: Economy and Space* 48, no. 4 (2016): 754–70.

Ibrahim, Nur Amali. *Improvisational Islam: Indonesian Youth in a Time of Possibility.* Ithaca NY: Cornell University Press, 2018.

"Jokowi Signs Regulation on Satu Data Indonesia." *Jakarta Post*, June 28, 2019. https://www.thejakartapost.com/news/2019/06/28/jokowi-signs-regulation-on-satu-data-indonesia.html.

Jong, Hans Nicholas. "Indonesia Calls on Oil Palm Industry, Obscured by Secrecy, to Remain Opaque." *Mongabay*, May 21, 2019. https://news.mongabay.com/2019/05/indonesia-calls-on-palm-oil-industry-obscured-by-secrecy-to-remain-opaque/.

Keyes, Os. "The Gardener's Vision of Data." *Real Life Mag*, May 6, 2019. https://reallifemag.com/the-gardeners-vision-of-data/.

Lee, Doreen. "Styling the Revolution: Masculinities, Youth, and Street Politics in Jakarta, Indonesia." *Journal of Urban History* 37, no. 6 (2011): 933–51.

Li, Tania Murray. *The Will to Improve: Governmentality, Development, and the Practice of Politics.* Durham NC: Duke University Press, 2007.

Lin, Cindy, and Silvia Margot Lindtner. "Techniques of Use: Confronting Value Systems of Productivity, Progress, and Usefulness in Computing and Design." *Proceedings of the 2021 CHI Conference on Human Factors in Computing Systems.* Article no. 595 (May 2021): 1–16.

Mazzarella, William, and Raminder Kaur. "Between Sedition and Seduction: Thinking Censorship in South Asia." In *Censorship in South Asia: Cultural Regulation from Sedition to Seduction*, edited by William Mazzarella and Raminder Kaur, 1–28. Bloomington: University of Indiana Press, 2009.

McQuillan, Dan. "Data Science as Machinic Neoplatonism." *Philosophy and Technology* 31, no. 2 (2018): 253–72.

Moon, Suzanne. "Justice, Geography, and Steel: Technology and National Identity in Indonesian Industrialization." *Osiris* 24, no. 1 (2009): 253–77.

Peluso, Nancy Lee. "Whose Woods Are These? Counter-mapping Forest Territories in Kalimantan, Indonesia." *Antipode* 27, no. 4 (1995): 383–406.

Setiadarma, Eunike Gloria. "Cultivating Pembangunan: Rice and the Intellectual History of Agricultural Development in Indonesia, 1945–65." Working paper, Northwestern University, Evanston IL, 2019. https://www.edgs.northwestern.edu/documents/working-papers/nike_egs_arryman-paper-draft_complete_04.pdf.

Shahab, Nabiha. *Indonesia: One Map Policy*. Open Government Partnership, December 2016. https://www.opengovpartnership.org/wp-content/uploads/2001/01/case -study_Indonesia_One-Map-Policy.pdf.

Smith, Brad. "A Healthy Society Requires a Healthy Planet." *Official Microsoft Blog*, April 15, 2020. https://blogs.microsoft.com/blog/2020/04/15/a-healthy-society-requires -a-healthy-planet/.

Strassler, Karen. *Demanding Images: Democracy, Mediation, and the Image-Event in Indonesia*. Durham NC: Duke University Press, 2020.

Taussig, Michael. *Defacement: Public Secrecy and the Labor of the Negative*. Stanford CA: Stanford University Press, 1999.

Thakor, Mitali. "Digital Apprehensions: Policing, Child Pornography, and the Algorithmic Management of Innocence." *Catalyst: Feminism, Theory, Technoscience* 4, no. 1 (2018): 1–16.

Tsing, Anna Lowenhaupt. *Friction: An Ethnography of Global Connection*. Princeton NJ: Princeton University Press, 2005.

Varkkey, Helena. "Patronage Politics as a Driver of Economic Regionalization: The Indonesian Oil Palm Sector and Transboundary Haze." *Asia Pacific Viewpoint* 53, no. 3 (2012): 314–29.

Conclusion

TOWARD A POLITICAL ECOLOGY OF DATA

Rebecca Lave, Eric Nost, and Jenny Goldstein

Political ecologists have a long track record of critiquing the politics and unjust consequences of environmental science. From the field's beginnings in the work of Piers Blaikie (1985), Susanna Hecht (1985), Michael Watts (1983), and Nancy Peluso (1992) to research in the mid-2000s from scholars including Diana Davis (2007), Becky Mansfield (2004), and Morgan Robertson (2006) to more recent work from Jessica Dempsey (2016), Rosemary Collard (2020), and Adam Romero (2021), among many others, political ecologists have examined how the theories, methods, and tools we use to understand the world around us profoundly shape how we manage it. But why should political ecologists pay attention to data infrastructures as they inform existing regimes of environmental governance and politics? Against ideas of data as immaterial or virtual, as driving decision-making, as resources with inherent value just waiting to be collected (data as the new oil or uranium), and as simultaneously more than adequate for but ineffective at social and environmental change, this book shows that data is material, governed, and practiced; requires praxis; and is very worthy of political-ecological critique.

The Nature of Data extends political ecologists' focus on knowledge production, circulation, and application and the role of experts and expertise in environmental management to digital and data technologies and names an emerging subfield, the *political ecology of data*.[1] The authors have adopted a broad range of methodologies and approaches, from collaborative modeling to ethnography, and addressed an equally broad range of empirical foci, from penguins to urban data centers. As their case studies have made clear, the abstraction of rich relational information into simplified, rationalized datasets is a complex socio-

303

technical endeavor. To paraphrase Dolly Parton, the central argument of this collection might be that it takes a lot of production to make data look this found.[2]

So what can one take away from this volume and from this emergent conversation on a political ecology of data? In the paragraphs that follow, we point to three larger questions and their sometimes ambiguous answers: What is data, if not the straightforward mirror of nature it is often imagined to be? What do data and infrastructure do in the world now? And what might they do in the future? We close with some suggestions for where a political ecology of data analysis might next turn its attention.

What Is the Nature of Data?

Drawing on existing scholarship from political ecology and science and technology studies (STS), the contributors to this edited volume show that we cannot fully understand the current conjuncture in global environmental politics without understanding the platforms, hardware devices such as sensors and servers, software such as algorithms and models, and institutions that enable actors to generate and analyze data. One approach authors in this volume draw from political ecology is to question the substance and context of environmental science and whether scientific data gets it right. This has been central to political ecologists' study of environmental knowledge regimes, from Blaikie's ignorant and detached "bureaucrat in the plane" flying over the communities for which he is responsible to calls for "socializing the pixel" that reduce complex land use processes to static categories.[3]

But beyond representing the material world (accurately or not), data does things within it. Building on traditions from political ecology and STS, the authors demonstrate that data is the hybrid offspring of technology and social practice. Data, even when it appears in its most simplified abstract form, is presented as the product of intimate, deeply affective social relations: among scientists themselves, between scientists and their field sites, and among environments, scientists, and the human and extrahuman communities that they call home (see chapters 3, 4, 6, 7, 9, and 14, among others). Finally, following the classic

Lave, Nost, Goldstein

STS playbook, the authors argue that social conditions of production lead to inherently social results. Multiple chapters argue that data is inherently political. Contrary to its commonly assigned attributes of objectivity and transparency, the authors show, data is socially produced and opaque. Data does not exist in isolation but only in social, political, and economic context. That makes the political economy of data deeply important: which data is made available (and which is not), who benefits from this data, and from whom it is extracted (chapter 2).

Data thus is performative, regardless of how accurately it captures social and environmental life. For instance, the creation of data infrastructure—the bringing together of seemingly objective information to rationally allocate resources—might legitimize or even destabilize a conservation program, regardless of how accurate, precise, or objective that data actually is (chapter 15). "Stronger" senses of performativity suggest that data—as "captures" of specific elements of a phenomenon and not others—tends to reproduce those elements over ones not captured. For example, some measures of hydropower's environmental footprint capture threatened species like salmon in accurate but binary, static, and discretized ways, potentially giving rise to a much different energy landscape than one in which the broader relations among species and between species and people are considered (chapter 14). The authors' shared critique of data as performatively neutral, but actually partial, leads into a far more diverse range of views on the ecosocial role of data today.

What Does Data Do Today?

What does the rise of big data accomplish in the present moment, particularly given that it is simultaneously characterized by what chapter 2 refers to as "continuous improvement and persistent limitations"? The volume introduces critical tensions in considering this but does not resolve them. While many of the authors view big data and its infrastructure as only promissory, others argue that big data is already here, reworking and disrupting current practice (on the former, see chapters 2, 3, 8, 9, 11, and 12; on the latter, see chapters 1, 3, and 10).[4] There is a similar lack of consensus as to whether big data is path dependent, promising continuity with the past, or instead a source of revolutionary change, though most authors take the former position (see chapters 1, 2, 3, 6, and 12).

Stemming from this, a third tension is whether big data is extractive, reaping commercial and political benefit and damaging people and environments, or is instead emancipatory, enabling more just environmental management. In the words of Eitzel and colleagues in chapter 6, big data and infrastructure "harm the environment (through limited understanding) and democracy (through limited access)." One common strand in this third set of tensions focuses on whether big data's promises of public access are realistic or if commercialization is not only inevitable but already happening (see chapters 1, 2, 10, 11, and 13). Another strand focuses on what the authors see as the inevitable biophysical dangers of big data, stemming primarily from the ways in which big data, as Corrine Armistead puts it in chapter 14, is distilled for "simplicity and efficiency rather than retaining richness" and relationality, and directly damages the environment through electricity demand (see chapter 4).

What Might Data Do in the Future?

As presented in the chapters of this volume, big-data enthusiasts put forward at least three competing visions for its future:

Global environmental omniscience, based on claims that ecological change is predictable, and therefore programmable into algorithms for processing abundant data

Objective, rational, and neutral governance based on claims that data is not arbitrary, wasteful, or partisan

Emancipation and decolonization through increased transparency and democracy

Many of the authors in this volume make the quintessential political ecological move of challenging enthusiasts' narratives, questioning the utopian visions of benign omniscience that enables better governance for humans and extra humans. These authors argue that the future of big data and the infrastructure behind it is actually one of support for authoritarian surveillance and extractivist practices, leading to the degradation of the environment and democracy. Other authors, while quite clear about the dangers, still seem to hope for a future more in keeping with enthusiasts' narratives: big data as support for different environmental governance, promising better outcomes for human and extrahuman communities suffering from environmental injustices. The chapters in this volume thus demonstrate the classic political ecology combination of incisive critique (hatchet) and productive response (seed).[5]

Moving Forward

The critical tensions surrounding big data's present and future brought into the foreground in this volume suggest that the political ecology of data is off to a good start. There is an appealing level of intellectual freedom in a body of work that can span "modeling of the oppressed" (chapter 6), "just good enough" citizen sensing (chapter 5), and penguin science as a tool of authoritarian populism (chapter 13). The suggestions that follow are thus intended to further open discussion on where a political ecology of data might go from here.

First, we should consider how to extend our thinking about data as relational. Authors in this volume carefully critique conventional definitions of data and show how data is a social product. And yet the authors often structure their critiques via binaries: data is either crisp or fuzzy, homogeneous or heterogeneous, official or lay, universalized and abstract or local and embedded. There is far less consideration of how data can be both. This binary thinking about data is particularly notable given that the authors define infrastructure in messier and more nuanced ways. Indeed, across the volume, infrastructure is rarely if ever portrayed as one thing or another. Instead, it is simultaneously technical, relational, and natural; both material and digital; abstracted but also shaped by the particularities of the landscape, resource, or ecosystem service being monitored. Even while functioning as a tool of governance, classification, and accumulation, infrastructure is granted complexity. A political ecology of data can do the same for the data it collects, sorts, and spits out again. For instance, data is abstract, but it is also felt and otherwise emotionally charged; recognizing this can shed light on how it enters into governance as contested (or not).

Second, the authors in this collection go beyond the references to Haraway, Jasanoff, and Latour that seem to have been sufficient shorthand for STS in political ecology since the mid-1990s. Instead, there are references to STS scholars such as Bowker, Fortun, Irwin, and Star. Political ecology of data could develop further by expanding its engagement with the STS community, which has a half-century head start on political ecology in thinking about the social character and implications of data and technology. Engaging with critical STS scholarship on the history of technology (e.g., Paul Edwards, Gabrielle Hecht, and Sara Pritchard), citizen science (e.g., Aya Kimura, Abby Kinchy, and Gwen Ottinger), and data sensing and measuring (e.g., Max Liboiron, Michelle Murphy, Nick Shapiro, and Sara Wylie) would both reshape what kinds of questions political ecologists ask about data and add empirical and conceptual depth.[6]

As the COVID-19 pandemic has made abundantly clear, we can no longer ignore data infrastructure if we are to understand the shape of contemporary politics. Infrastructures for tracking cases, depicting trends, and informing policy decisions reflected the unevenness of public health systems across scales and became evidential pivot points for those crying "hoax" as well as those pointing out disproportionate harm to Black, Indigenous, and people of color. Drawing on the insights of this volume and the critical tensions the authors raise, and picking up on these potential paths forward, political ecologists will be well positioned to make sense of, question, and recraft emerging infrastructures that log and depict environmental and climate crises to inform environmental governance.

Notes

1. Forsyth, *Critical Political Ecology*; Goldman, Nadasdy, and Turner, *Knowing Nature*; Lave, "Bridging Political Ecology and STS"; Lave, "Engaging within the Academy."

2. "It takes a lot of time and money to look this cheap." Dolly Parton, Twitter post, July 2011, 8:43 p.m.

3. Blaikie, *Political Economy*; Geoghegan et al., "'Socializing the Pixel.'"

4. In chapter 3 Bakker and Ritts argue both that there are already effects from the rise of Smart Earth technologies and that more are to come.

5. Robbins, "Trickster Science."

6. Much can also be learned from pioneering work in participatory modeling in geography, such as work from Stuart Lane and Rachel Pain.

Bibliography

Blaikie, Piers M. *The Political Economy of Soil Erosion in Developing Countries*. London: Longman, 1985.

Collard, Rosemary. *Animal Traffic: Lively Capital in the Exotic Pet Trade*. Durham NC: Duke University Press, 2020.

Davis, Diana K. *Resurrecting the Granary of Rome: Environmental History and French Colonial Expansion in North Africa*. Athens: Ohio University Press, 2007.

Dempsey, Jessica. *Enterprising Nature: Economics, Markets, and Finance in Global Biodiversity Politics*. Chichester, UK: Wiley-Blackwell, 2016.

Forsyth, Timothy. *Critical Political Ecology: The Politics of Environmental Science*. London: Routledge, 2003.

Geoghegan, Jacqueline, Lowell Pritchard Jr., Yelena Ogneva-Himmelberger, Rinku Roy Chowdhury, Steven Sanderson, and B. L. Turner II. "'Socializing the Pixel' and 'Pixelizing the Social' in Land-Use and Land-Cover Change." In *People and Pixels: Linking Remote Sensing and Social Science*, edited by National Research Council, 51–69. Washington DC: National Academies Press, 1998.

Goldman, Mara J., Paul Nadasdy, and Matthew D. Turner. *Knowing Nature: Conversations at the Intersection of Political Ecology and Science Studies*. Chicago: University of Chicago Press, 2011.

Hecht, Susanna. "Environment, Development and Politics: Capital Accumulation and the Livestock Sector in Eastern Amazonia." *World Development* 13, no. 6 (1985): 663–84.

Lave, Rebecca. "Bridging Political Ecology and STS: A Field Analysis of the Rosgen Wars." *Annals of the Association of American Geographers* 102, no. 2 (2012): 366–82.

——. "Engaging within the Academy: A Call for Critical Physical Geography." *ACME: An International E-Journal for Critical Geographies* 13, no. 4 (2014): 508–15.

Mansfield, Becky. "Rules of Privatization: Contradictions in Neoliberal Regulation of North Pacific Fisheries." *Annals of the Association of American Geographers* 94, no. 3 (2004): 565–84.

Peluso, Nancy Lee. *Rich Forests, Poor People: Resource Control and Resistance in Java*. Berkeley: University of California Press, 1992.

Robbins, Paul. "The Trickster Science." In *The Routledge Handbook of Political Ecology*, edited by Tom Perreault, Gavin Bridge, and James McCarthy, 89–101. New York: Routledge, 2015.

Robertson, Morgan M. "The Nature That Capital Can See: Science, State and Market in the Commodification of Ecosystem Services." *Environment and Planning D: Society and Space* 24, no. 3 (2006): 367–87.

Romero, Adam. *Economic Poisoning: Industrial Waste and the Chemicalization of American Agriculture*. Berkeley: University of California Press, 2021.

Watts, Michael. *Silent Violence: Food, Famine, & Peasantry in Northern Nigeria*. Berkeley: University of California Press, 1983.

SOURCE ACKNOWLEDGMENTS

Portions of chapter 1 were previously published in "The Factories of the Past Are Turning into the Data Centers of the Future," *The Conversation*, January 3, 2017.

Portions of chapter 3 were previously published in "Smart Earth: A Meta-Review and Implications for Environmental Governance," *Global Environmental Change* 52 (September 2018): 201–11.

A version of chapter 5 was previously published as "Just Good Enough Data and Environmental Sensing: Moving beyond Regulatory Benchmarks toward Citizen Action," *International Journal of Spatial Data Infrastructures Research* 13 (2018): 4–14.

A version of chapter 10 was previously published as "Practicing Environmental Data Justice: From DataRescue to Data Together," *Geo: Geography and Environment* 5, no. 2 (2018): e00061.

Portions of chapter 13 were previously published in "Tracking Penguins, Sensing Petroleum: 'Data Gaps' and the Politics of Marine Ecology in the South Atlantic," *Environment and Planning E: Nature and Space*. © 2019 by the author. Reprinted by permission of SAGE Publications, Ltd.

A version of chapter 15 was previously published as "How to Make a Forest," *At the Border* (e-flux Architecture and A/D/O), April 10, 2020. Used with permission.

CONTRIBUTORS

Luis F. Alvarez León is an assistant professor of geography at Dartmouth College and the director of the Critical Geospatial Analysis Lab. He is a political economic geographer with substantive interests in geospatial data, media, and technologies. His work integrates the geographic, political, and regulatory dimensions of digital economies under capitalism, with an emphasis on technologies that manage, represent, navigate, and commodify space.

Corrine Armistead is a PhD student in the Department of Geography at the University of British Columbia. She holds an MS in geospatial technologies from the University of Washington, Tacoma, and her research incorporates analysis of GIS and remote-sensing technologies in water and energy governance.

Karen Bakker is a professor of geography at the University of British Columbia. Her recent work, at the intersection of political economy and political ecology, focuses on the implications of novel digital technologies for environmental governance. She is the author of over one hundred academic publications and sits on the boards of the International Institute of Sustainable Development and the UN Internet Governance Forum's Policy Network on Environment (Multistakeholder Working Group).

James J. A. Blair is an assistant professor of geography and anthropology at California State Polytechnic University, Pomona. Blair holds a PhD in anthropology from the Graduate Center, City University of New York. His research on sovereignty, natural resources, and environmental science is supported by the National Science Foundation, Fulbright-

IIE, Wenner-Gren Foundation, Social Science Research Council, and Mellon Foundation/American Council of Learned Societies.

Irus Braverman is a professor of law and adjunct professor in geography at the University at Buffalo, the State University of New York. Her books include *Planted Flags: Trees, Land, and Law in Israel/Palestine* (2009), *Zooland: The Institution of Captivity* (2012), *Wild Life: The Institution of Nature* (2015), *Coral Whisperers: Scientists on the Brink* (2018), and *Zoo Veterinarians: Governing Care on a Diseased Planet* (2021). Braverman is currently writing about nature conservation, animality, and settler colonialism in Palestine/Israel.

Abraham Changarara is a member of the Muonde Trust Digital Documentation team. His research interests include digital documentation and mapping of sacred places, community institutions and resources, social service delivery, and governmental and nongovernmental interventions at the margins.

Adnomore Chirindira is the agro-ecological officer at Muonde Trust. He is well versed in community-managed forests, rivers, mountains, and wildlife. His research interests are in zero-chemical projects and wildlife management.

Lindsey Dillon is an assistant professor in the Department of Sociology at the University of California, Santa Cruz, and is affiliated with the Environmental Studies Department, the Community Studies Program, and the Science and Justice Center. She studies environmental and economic justice in U.S. cities.

M. V. Eitzel is a researcher at the University of California, Santa Cruz, in the Science and Justice Research Center. She holds a PhD in environmental science, policy, and management from the University of California, Berkeley. Dr. Eitzel's research interests include synthesizing multidecadal legacy natural history datasets, critical participatory data science, community-based modeling, and complex social-ecological system management.

Madeleine Fairbairn is an associate professor in the Environmental Studies Department at the University of California, Santa Cruz. She studies the political ecology of agri-food systems through the lenses of sociology and geography, including research on finance-sector investment in farmland and the emerging Silicon Valley agri-food tech sector. Her first book, *Fields of Gold: Financing the Global Land Rush* (2020), is available as an open-access ebook.

Hilary O. Faxon is a human geographer whose research examines how digital tools extend and transform political ecologies, agrarian relations, and global development. She held a Ciriacy-Wantrup Postdoctoral Fellowship at the University of California, Berkeley, and is currently a Marie Sklodowska-Curie Fellow at the University of Copenhagen. She has worked in Myanmar since 2014.

Aaron C. Fisher is a computational scientist and head of the Numerical Analysis and Simulations Group at Lawrence Livermore National Laboratory. His research interests include parallel computing, finite element methods, and using simulation for more efficient industrial practices.

Jennifer Gabrys is the chair in media, culture, and environment in the Department of Sociology at the University of Cambridge. She leads the Planetary Praxis research group and is the principal investigator of Smart Forests: Transforming Environments into Social-Political Technologies. She is the author of *How to Do Things with Sensors* (2019), *Program Earth: Environmental Sensing Technology and the Making of a Computational Planet* (2016), and *Digital Rubbish: A Natural History of Electronics* (2011).

Patrick Gallagher is an assistant professor in the Department of Anthropology at the University of Texas at San Antonio. His work examines the cultural politics of nature through ethnographic fieldwork with market-oriented conservation projects in coastal Belize and the United States.

Jenny Goldstein is a geographer and assistant professor in the Department of Global Development at Cornell University. She is also core faculty in Cornell's Southeast Asia Program and was previously an Atkinson Center for Sustainability Postdoctoral Associate in the Department of Science and Technology Studies at Cornell. Her research interests include the relationships among digital technology, environmental governance and biophysical change, as well as the politics of knowledge and land use in Indonesia.

Zbigniew Grabowski works as a postdoctoral research associate at Cary Institute of Ecosystem Studies, the Urban Systems Lab at the New School, and an adjunct assistant professor at Portland State University, where he teaches classes in anthropology and geography. His work focuses on the intersection of sociotechnical systems evolution, biocultural conservation, and justice. He has research interests in green infrastructure, river restoration, dam removal, experiential environmental learning, and governance.

Emmanuel Mhike Hove is the arts, culture, and education officer at the Muonde Trust. He holds a first-class B s c honors degree in music and musicology and is currently studying toward a master of arts in development studies. His research interests revolve around biocultural resilience, building on traditional ecological knowledge for community resilience and sustainability.

Noor Johnson is a research scientist and environmental anthropologist at the National Snow and Ice Data Center, University of Colorado Boulder, where she leads the Exchange for Local Observations and Knowledge of the Arctic. Her research has focused on Arctic environmental governance, Indigenous food security and sovereignty, equity in climate change research and policy, and community-based monitoring. She has worked with Indigenous, international, and research organizations on issues related to science and environmental policy.

Zenia Kish is an assistant professor of media studies at the University of Tulsa. Her work explores global media, sociotechnical imaginaries, infrastructures of food and agriculture, and philanthrocapitalism. She is writing a book on philanthropic media culture, *The Invisible Heart of Markets: Philanthrocapitalism and the Mediation of Development*, and is coeditor with Emily Contois of the anthology *#FoodInstagram: Identity, Influence & Negotiation*. She is a member of the Agri-Food Technology Research Project at the University of California, Santa Cruz.

Rebecca Lave is a professor and the chair of geography at Indiana University. Her research combines physical and social science and focuses on the construction of scientific expertise, market-based environmental management, and water regulation in the United States.

Aaron Lemelin is a member of the Environmental Data and Governance Initiative's web monitoring team and has previously worked for the Sunlight Foundation's Web Integrity Project.

Anthony Levenda is an assistant professor in the Department of Geography and Environmental Sustainability at the University of Oklahoma. Currently, his research explores policy and activism for just energy transitions and the urban politics of climate adaptation. Dr. Levenda has long-standing interests in the environmental implications of large-scale data infrastructure, the growing movements for climate justice, and the overlapping concerns of energy justice and climate adaptation in the U.S. Gulf Coast region.

Cindy Lin is an assistant professor in the College of Information Sciences and Technology at Pennsylvania State University. She held a postdoctoral fellowship at the Cornell Atkinson Center for Sustainability and the Department of Information Science. Her research investigates the presence of bias and inequality in artificial intelligence (AI) systems deployed within the environmental sciences. She is a coauthor of the multigraph *Technoprecarious* (2020).

Abraham Mawere Ndlovu is the executive director at the Muonde Trust. He has spent more than three decades working in a unique kind of community action research revolving around human welfare and ecological dynamics. His research interests include community-driven development, permaculture, Indigenous knowledge systems, agropastoral management, community land, and Zimbabwe Liberation War history.

Alice Ndlovu is the director of operations and administration at the Muonde Trust. She has honors and a master's degree in development studies from Midlands State University. In addition, she holds a certificate in "working with communities affected by poverty displacement and HIV and AIDS" from the University of Kwazulu Natal in South Africa.

Daniel Ndlovu is the research officer at the Muonde Trust. He holds a BSc honors degree in geography and environmental studies. His research interests include land history and environmental and community sustainability.

Kleber Neves is a researcher with the Brazilian Reproducibility Initiative at Federal University of Rio de Janeiro. He has a PhD in evolutionary neuroscience, and his research interests include applications of computational approaches to neuroscience and metascience.

Eric Nost is an assistant professor in the Department of Geography, Environment and Geomatics at University of Guelph and a member of the Environmental Data and Governance Initiative. He researches how data technologies inform environmental governance and conservation, drawing on the fields of political ecology, science and technology studies, and digital geographies.

Oluwasola E. Omoju is a research fellow in the Department of Economic and Social Research of the National Institute for Legislative and Democratic Studies, Abuja, Nigeria. He received his PhD in applied economics (energy economics) from Xiamen University, China. His research interests include energy economics and policy, public financial management, sustainable development (with special reference to Africa), and policy impact evaluation using computable general equilibrium modeling and applied microeconometrics (experimental and quasi-experimental techniques).

Graham Pickren is an assistant professor of sustainability studies at Roosevelt University, Chicago. His areas of expertise converge around three interrelated themes: an urban political ecology approach to the study of cities and nature; an interest in green political economy and debates about sustainability, particularly a Green New Deal; and the development of renewable energy infrastructure that is publicly owned.

Helen Pritchard is an associate professor in queer feminist technoscience and digital design at i-DAT, University of Plymouth, where she is also the program lead for MRes Digital Art and Technology. Helen's work considers the impacts of computation on social and environmental justice. She is coeditor of *DataBrowser 06: Executing Practices* (2018) and the special issue "Sensors and Sensing Practices" in *Science, Technology and Human Values* (2019).

Peter Pulsifer is an associate professor with the Department of Geography and Environmental Studies and the associate director of the Geomatics and Cartographic Research Centre, Carleton University, Ottawa, Canada. His research addresses questions around computer-based information representation, with a particular focus on interoperability and data sharing across knowledge domains. He has extensive experience working with Indigenous communities in a codesign and coproduction model for the establishment of community-based expertise, capacity, and information systems.

Max Ritts is a postdoctoral fellow at the University of Cambridge. His work operates at the intersection of political ecology, sound studies, and critical Indigenous studies, with a particular focus on environmental governance. He is the author of the forthcoming monograph *A Resonant Ecology*.

Jon Solera founded Seven Points Consulting, an engineering management consultancy specializing in software development. He received his bachelor's degree in electrical engineering from Princeton and pursued graduate studies at Stanford. After over a decade at Hewlett-Packard and Google, he worked at several small companies before going into consulting. His client base includes the public and private sectors as well as NGOs.

Colleen Strawhacker is a program director in the Arctic Sciences Section in the Office of Polar Programs at the National Science Foundation. She holds a PhD in anthropology from Arizona State University, where she focused on the archaeology of climate-driven risk and vulnerability to food security in the U.S. Southwest and the North Atlantic. Her expertise includes research approaches spanning the natural and social sciences and developing cyberinfrastructure for the social sciences and Indigenous knowledge.

André Veski is a senior data scientist at Starship Technologies. He received his PhD in informatics from Tallinn University of Technology, Estonia. His research interests include market design, matching markets, social interaction simulation models, and strategic thinking.

Dawn Walker is a PhD candidate at the Faculty of Information, University of Toronto. Her research focuses on the values, transformations, and imagined futures in the design of decentralized alternatives to existing digital infrastructures.

K. B. Wilson trained in zoology and anthropology at the Universities of Oxford and London. He now lives in Sabah, North Borneo, having retired from a career in philanthropy and academia focused on biocultural diversity and backing Indigenous territorial stewardship around the world. He first lived in Zimbabwe as a schoolteacher in 1980 and has been engaged in action research in Mazvihwa Communal Area alongside the Muonde Trust, a local NGO, ever since.

INDEX

Facebook (*cont.*)
  as platform for activists, 97, 147, 238, 242;
    and surveillance, 43, 52, 239
Falkland Islands (Malvinas), 251–57, 261, 262
FarmDrive, 212, 221
farmers, 214, 215, 222–24; in California, 1; and
    connectivity, 217; and the "data deficit," 214–
    16; and the data revolution in agriculture,
    211–13; in the Global South, 2, 10, 179, 185;
    smallholder, 211, 213, 216–20, 237; women, 123
Farmforce, 222
Federal Records Act, 201
fiber-optic cables, 21–24, 27, 29, 31, 33, 65, 83,
    87, 202
finance capital, 28, 29
"financialization of nature," 28, 29, 33
financial services, 22, 212, 221
fishing, 144, 156, 231; and the Columbia River
    Basin, 87–91; and the Falklands, 251–54, 258;
    illegal, 72; Indigenous practices of, 93, 94;
    and sovereignty, 251
Fitbit, 64, 72
fixed sensors, 63
Florida, 140, 142, 144
forecasting, 71
forests, 64, 71, 72, 131, 176, 185; and commodi-
    fication, 21, 29; and data, 1, 2, 6, 62, 285; fires
    in, 71, 297; in Indonesia, 287–93, 296–99;
    and maps, 11, 270, 287; in Myanmar, 239, 241;
    and satellites, 42, 179–81; in Southeast Asia,
    230, 233–35. *See also* deforestation
Frackbox, 110. *See also* Citizen Sense Kit
fracking, 8, 107–12
"fuzzy" classification, 183–85, 307

Gap Project, 257, 259–61
gemstone mining, 240, 243
genes, 89, 138, 154
Geodata for Agriculture and Water (Nether-
    lands), 216
geo-fencing, 63
Geographic Information Systems (GIS): and data
    structures, 268, 269; and ground-truthing, 174–
    76; and knowledge production, 4; and open-
    data initiatives, 243; and pixels, 184–87, 234; and
    QGIS, 125; and salmon tracking, 278, 279, 281
Ghana, 221, 222

glaciers, 154
Global Earth Observing System of Systems
    (GEOSS), 155
Global Fishing Watch, 72
Global Forest Watch, 72, 234
global location sensing (GLS) tags, 257, 258
Global Open Data for Agriculture and Nutri-
    tion (GODAN), 218
Global Positioning System (GPS), 63, 211, 215,
    235, 257, 258, 292, 295, 298
Golden Triangle, 234
Google, 1, 4, 25, 43, 52, 96, 97
Google Earth, 62, 241
governance, 2, 4, 7, 76, 107, 155, 218, 223, 306;
    community-scale, 166; conservation, 174,
    175; data-driven, 197; and the Environmen-
    tal Data and Governance Initiative, 9, 191;
    global, 74; and Indigenous sovereignty,
    157; natural resource, 10, 11, 237, 240, 243;
    predictive, 71, 72; reforms, 73; regimes, 233,
    244; and settler colonialism, 91; and Smart
    Earth, 65, 66, 75; sustainable, 262; and
    technology, 68
Grameen Foundation, 211, 221
graph databases, 11, 267–69, 276, 277–81
Great Barrier Reef, 137, 139, 144
greenhouse gas emissions, 95, 201
Green New Deal, 192
Green Revolution, 10, 214, 215, 220
"ground-truthing," 139, 174–79, 182–87, 242
Group on Earth Observations (GEO), 154
Guerilla Archiving, 192, 193

habitats, 92, 97; and citizen science, 1, 137, 144;
    and conservation, 3, 175; salmon, 93, 94, 97,
    278–80; seabird, 261
hedge funds, 28, 32
Hiber, 43
high-frequency trading (HFT), 22, 28, 29
history, 7, 220, 237, 239, 252; of agricultural
    development, 211, 223; and historical period-
    ization, 83; of imperialism, 50, 255; local, 168,
    215; and place, 6; and predictive analysis, 72;
    of technology, 308
Hutanwatch, 234
hydraulic fracturing. *See* fracking
hydroelectric industry, 83, 95, 96

hydropower, 92–96; and the Columbia River Basin, 267, 268, 271–78, 280, 281; and resource governance regimes, 11, 84, 86; and threatened species, 305. *See also* dams

IBM, 62, 66
Indigenous peoples, 83, 89, 90, 308; in Argentina, 256; in Cambodia, 235; and data systems, 89; and fishing, 93, 94, 95; in Indonesia, 296; and knowledge systems, 84–87, 92, 120, 121, 125, 154–59, 163–70; lifeways of, 98; and observation networks, 161, 162; ontologies of, 11, 267, 273, 277, 279; and politics, 91, 97, and resource management, 120, 150; and science projects, 9; and societies, 88; sovereignty of, 7, 96; in Zimbabwe, 123
Indonesia, 6, 11, 46, 55, 64, 234, 287, 295, 299; and development, 288; and ground-truthing, 285, 296, 298; National Mapping Agency in, 289, 290, 291, 292; New Order Regime in, 297
infrastructure, 23–25, 65, 83, 84, 114, 202, 277; and citizen sensing, 113; and conservation, 178, 186; of dams, 93, 94; definition of, 176, 177; digital, 139, 156, 285; of energy, 95–97, 255, 274; of environmental sensing, 105; and infrastructuring, 157, 198; of logistics, 217; modeling, 119, 121; natural, 174; and natural gas, 107, 108, 110–12; of rail, 92; regional, 98; of roads, 276; social, 128, 130, 132, 203; social life of, 175; technical, 122, 125–27, 133, 192
insect biotelemetry, 63
International Union for Conservation of Nature, 144
internet, 1, 28, 87; in Africa, 127; and decentralization, 202; and development, 23; and digital preservation, 195, 198, 199, 200; and Google, 52; and the Internet Archive, 193, 194; and the "Internet of Things," 21, 43; and Microsoft, 285; in Southeast Asia, 230, 236–39
interoperable technologies, 67
Inuit-led organizations, 157, 163, 164, 167
investment, 121, 154, 243, 270; and energy markets, 95; federal, 97; and financialization, 29; foreign direct, 238, 289; and real estate, 24; and technology, 4, 23; and transparency, 2
irrigation, 87, 92, 93, 242
Italy, 49, 54–59

jade, 233, 237, 239–44
Jakarta, 293, 295
Japan, 39, 48, 49, 55, 57, 155
Java, 125, 288

Kalimantan, 286, 288, 291–93
Kenya, 38, 49–52
kin relations, 90
Kirchner, Cristina Fernández de, 255
Kirchner, Néstor, 252
Knight Lab (Northwestern University), 241

labor, 85, 187; and collaborative modeling, 120; of conservation, 175, 176; and data, 242, 274, 291, 296; digital, 29; and exploitation, 231, 240; of ground-truthing, 186, 187; and infrastructure, 21, 93, 202; of mapmaking, 11
Landsat, 37, 39, 40, 42, 179–81, 188n19
Laos, 46, 55, 234, 243
latency, 22, 30–32
lidar (light detection and ranging) data, 292–94
listening stations, 65
logging, 10, 72, 93, 231–36, 239, 292, 295
logistics, 96, 217
London Carbon Exchange, 33
Louisiana, 1, 196

Malawi, 2, 214
Malaysia, 55, 234
maps, 52, 64; and agriculture, 213, 215, 222; and the Arctic, 158, 160; and the Columbia River Basin, 267, 268, 270–72, 276, 277, 281; and dam development, 86, 279; and digital mapping, 125, 131; and environmental governance, 66; and Indonesia, 285, 286–98; and Landsat imaging, 180–83, 187; and modeling, 179; and Myanmar, 233, 234, 239, 240, 242; and natural infrastructure, 174, 175, 177; and OakMapper, 72; and participatory mapping, 119, 128; and situational mapping, 122
Marcellus Shale region, 108
markets, 148; and agriculture, 211–13, 220, 222, 223; and colonialism, 46; commodity, 21, 22; crashes of, 71; and the digital economy, 43–45, 66; and the energy sector, 84, 95, 96, 257; financial, 221; and fur, 92; and gemstones, 240; and "green" capitalism, 28; and infrastructure,

markets (*cont.*)

176; and market-oriented conservation, 187; and measurement, 280; and neoliberalism, 62, 174, 175, 230, 281; stock, 30, 32, 33; and volatility, 217

Mazvihwa Communal Area (Zimbabwe), 121, 122, 123

measurement, 44, 66, 69, 71, 75, 106, 107, 114, 126, 139, 267, 269, 280, 281, 286

memory, 83, 87, 252

microlending, 221

Microsoft, 4, 25, 62, 65, 285

microwave towers, 23, 31, 32

mining industry, 1, 25, 92, 231, 234, 237, 240–42, 259

mobile apps, 63

modeling, 119, 121, 179, 182; agent-based, 9, 125; collaborative, 8, 120, 123, 124, 132, 133, 303; and infrastructure, 125, 128, 177; labor of, 175; and the material, 176; and "modeling of the oppressed," 120, 307; process of, 122, 130, 131

Morocco, 46, 55

Muonde Trust (Zimbabwe), 121–23, 127, 128, 130, 133

Myanmar, 10, 230, 233, 234, 237–44

Nairobi, 48, 221

NASA, 41, 50, 201

Nasdaq, 31

National Hydropower Asset Assessment Program (NHAAP), 271

National Oceanic and Atmospheric Administration (NOAA), 139

National Snow and Ice Data Center (NSIDC), 157

natural capital, 33, 175–77

natural resources, 1, 62, 74, 231, 237, 240, 252, 256, 261

nature, 7, 12, 98, 137, 181; as capital, 21, 23, 28, 178, 252; and climate change, 119, 216; and colonialism, 83; and conservation, 138, 186; and data technologies, 3, 84–89, 97, 267–70, 285, 304; digitization of, 1, 4; and economic value, 176, 186; and ecosystem services, 174; financialization of, 29–34; and human-nature relations, 87, 88, 92; and infrastructure, 93, 105; models of, 175; and race and class, 287;

representations of, 177, 185, 187, 271, 272, 276, 281, 299; and society, 11; visualization of, 177

Nature Conservancy, 1

neoliberalism, 7, 28, 33, 38, 174, 175, 224

NetLogo, 124–26, 129, 130, 132

Nigeria, 38, 46, 56

nongovernmental organizations (NGOs), 2, 10, 63; and agriculture, 212, 213, 218, 223; in Myanmar, 239, 240, 242; and predictive algorithms, 119; and wildlife preservation, 236; in Zimbabwe, 121

North American Citizen Science Association, 113

OakMapper, 72

observatories, 64, 68, 162

oil industry, 11, 64, 74, 109, 237; and environmental review, 257, 259, 260; in the Falklands, 251, 252, 253, 256; and offshore drilling, 254, 262; and oil frontiers, 255; and oil spills, 251, 258. *See also* palm oil

One Data, 11, 285–86

One Map, 11, 239, 242, 285, 286, 289, 290

Ooredoo (telecom company), 238

opium, 234, 237

Pacific Northwest (U.S.), 25, 88; and data infrastructures, 83, 87; and hydropower, 11, 93, 267, 271–73; Indigenous communities of, 7, 88, 277, 279. *See also* Columbia River Basin (CRB)

palm oil, 286, 288–90, 296

Pampa Azul (Blue Pampa) campaign (Argentina), 255, 256, 260

participatory mapping, 119, 128

peer-to-peer technologies, 199, 202

penguins, 25–53, 257, 258, 260, 262, 303

Pennsylvania, 31, 107–9, 112, 115

Pennsylvania Department of Environmental Protection (DEP), 108

performativity, 305

petroleum. *See* oil industry

pixels, 178–82, 183–87, 234, 288, 291–96, 304

Planet (satellite start-up), 1, 7, 41–44, 52

Planetary Computer, 285, 286. *See also* Microsoft

poaching, 10, 63, 72, 231, 233, 234, 236

political ecology, 12, 87, 307, 308; of data, 3, 5, 233, 237, 244, 254, 286, 299, 303, 304; definition of, 6; and expertise, 7

Index

9 781496 217158